ILLUSTRATED PHYSIOLOGY

CHURCHILL LIVINGSTONE
Medical Division of Longman Group UK Limited

Distributed in the United States of America by
Churchill Livingstone Inc., 1560 Broadway,
New York, N.Y. 10036, and by associated
companies, branches and representatives
throughout the world.

First Edition 1963
    Italian translation 1966
Second Edition 1970
    Danish translation 1973
Third Edition 1975
    Japanese translation 1976
    Spanish translation 1981
Fourth Edition 1983
    Reprinted 1984
    Reprinted 1986

ISBN 0-443-02713-7

Produced by Longman Group (FE) Ltd
Printed in Hong Kong

ILLUSTRATED PHYSIOLOGY

BY

ANN B. McNAUGHT

M.B., Ch.B., Ph.D.
The Institute of Physiology, The University of Glasgow

AND

ROBIN CALLANDER

FFPh FMAA AIMBI
Director, Medical Illustration Unit, The University of Glasgow

FOURTH EDITION

CHURCHILL LIVINGSTONE

EDINBURGH LONDON AND NEW YORK

1983

THIS BOOK
IS
AFFECTIONATELY
DEDICATED

To my husband, James A. Gilmour

Ann B. McNaught

To my wife, Elizabeth

Robin Callander

PREFACE TO THE FOURTH EDITION

We continue to be encouraged by the favourable reception accorded *"Illustrated Physiology"* and are again indebted to reviewers and colleagues for suggested improvements. The present edition updates the material on the control of arterioles in different parts of the body and includes new material on intestinal secretions, haemostasis, blood coagulation and the reproductive system.

1983   A.B. McN.
                                                                            R.C.

PREFACE TO THE FIRST EDITION

This book grew originally from the need to provide visual aids for the large number of students in this department who come to the study of human physiology with no background of mammalian anatomy and often without any conventional training in either the biological or the physical and chemical sciences. Such students include postgraduates studying for the Diploma or Degree in Education; undergraduates working for Degrees in Science; laboratory technicians taking courses for the Ordinary and Higher National Certificates in Biology; and the increasing number of medical auxiliaries — physiotherapists, occupational therapists, radiographers, cardiographers, dietitians, almoners and social workers — many of whom are required to study to quite advanced levels at least regional parts of the subject.

Many medical, dental and pharmacy students, as well as nurses in training, have been good enough to indicate that they too find diagrams which summarize the salient points of each topic valuable aids to learning or revision. It is our hope that some of these groups may find our book helpful.

Each page is complete in itself and has been designed to oppose its neighbour. It is hoped that this will facilitate the choice of those pages thought suitable for any one course while making it easy to omit those which are too detailed for immediate consideration.

A book of this sort is largely derivative and it is impossible to acknowledge our wider debt. We wish to record our gratitude, however, to Professor R.C. Garry for his generous permission to borrow freely from the large collection of teaching diagrams built up over the years by himself and his staff; to Dr H.S.D. Garven and Dr G. Leaf for permission to use some of their own teaching material; and to Messrs Ciba Pharmaceutical Products Inc. from whose fine book of Medical Illustrations by Dr Frank H. Netter the diagram of the Cranial Nerves has been modified.

We are indebted to the following colleagues and friends who read parts of the original draft and offered helpful criticism:— Dr H.S.D. Garven, Dr J.S. Gillespie, Mr J.A. Gilmour, Dr M. Holmes, Dr B.R. Mackenna, Mr T. McClurg Anderson, Dr I.A. Boyd, Dr R.Y. Thomson, Dr J.B. deV. Weir.

We should like to express our gratitude to Mr Charles Macmillan and Mr James Parker of Messrs. E. & S. Livingstone Ltd. for their unfailing courtesy and encouragement, and to Mrs Elizabeth Callander for help in preparing the index.

January, 1963

Ann B. McNaught
Robin Callander
Institute of Physiology,
The University of Glasgow.

CONTENTS

CONTENTS

CONTENTS

CHAPTER 1

INTRODUCTION:

The TISSUES

The AMOEBA

All living things are made of PROTOPLASM. Protoplasm exists in MICROSCOPIC UNITS called CELLS. The SIMPLEST living creatures consist of ONE CELL. The AMOEBA (which lives in pond water) exemplifies the BASIC STRUCTURE of all animal cells and shows the PHENOMENA which distinguish living from non-living things.

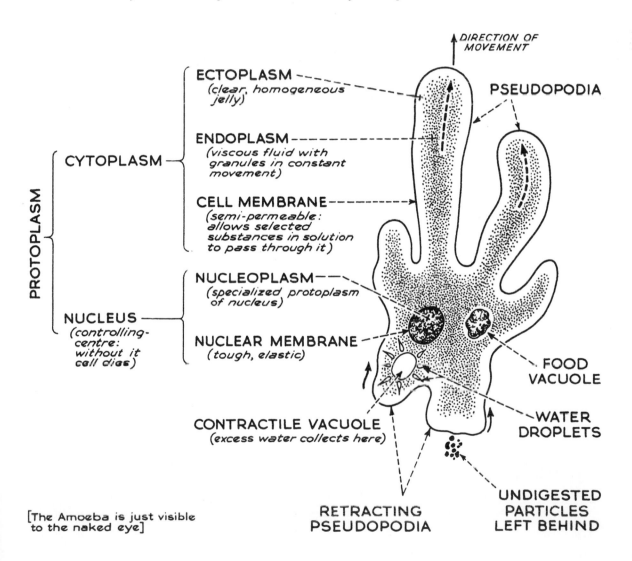

DIRECTION OF MOVEMENT

PROTOPLASM

CYTOPLASM

ECTOPLASM
(clear, homogeneous jelly)

ENDOPLASM
(viscous fluid with granules in constant movement)

CELL MEMBRANE
(semi-permeable: allows selected substances in solution to pass through it)

NUCLEUS
(controlling-centre: without it cell dies)

NUCLEOPLASM
(specialized protoplasm of nucleus)

NUCLEAR MEMBRANE
(tough, elastic)

PSEUDOPODIA

FOOD VACUOLE

WATER DROPLETS

CONTRACTILE VACUOLE
(excess water collects here)

RETRACTING PSEUDOPODIA

UNDIGESTED PARTICLES LEFT BEHIND

[The Amoeba is just visible to the naked eye]

3

The PHENOMENA which characterize all living things ----- are shown by the AMOEBA

① ORGANIZATION
Autoregulation —
inherent ability to control
all life processes.

② IRRITABILITY
Ability to respond to
stimuli (from changes
in the environment).

③ CONTRACTILITY
Ability to move.

④ NUTRITION
Ability to ingest,
digest, absorb
and assimilate
food.

⑤ METABOLISM and GROWTH
Ability to liberate potential
energy of food and to convert
it into mechanical work
(*e.g. movement*) and to
rebuild simple absorbed
units into the complex
protoplasm of the living cell.

⑥ RESPIRATION
Ability to take in oxygen
for oxidation of food
with release of energy;
and to eliminate the
resulting carbon
dioxide.

⑦ EXCRETION
Ability to eliminate
waste products of
metabolism.

⑧ REPRODUCTION
Ability to reproduce the
species.

PSEUDOPODIA FORMATION

② ③

Small elevation arises on surface

Cytoplasm streams forward

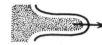
forming long pseudopodium

Free floating   Contact made   Crawling along surface

Attracted by favourable stimuli.
Repelled by unfavourable stimuli.

FOOD VACUOLE FORMATION

④

INGESTION: *Amoeba flows round and engulfs food particle.*

SECRETION of Enzymes (*chemical agents*) which DIGEST (*break down*) complex foods.

ABSORPTION & UTILIZATION of simple units by the living cell.

EXPULSION of undigested particles.

RESPIRATION and EXCRETION

⑥ ⑦

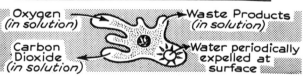

Oxygen (*in solution*)
Carbon Dioxide (*in solution*)
Waste Products (*in solution*)
Water periodically expelled at surface

REPRODUCTION   Asexual in the amoeba — by simple fission.

⑧

The PARAMECIUM

The Paramecium (another ONE-CELLED fresh water creature) shows:—

MODIFICATION and localization ----- for ----- SPECIALIZATION and
        of STRUCTURE   localization of
                                              CERTAIN FUNCTIONS

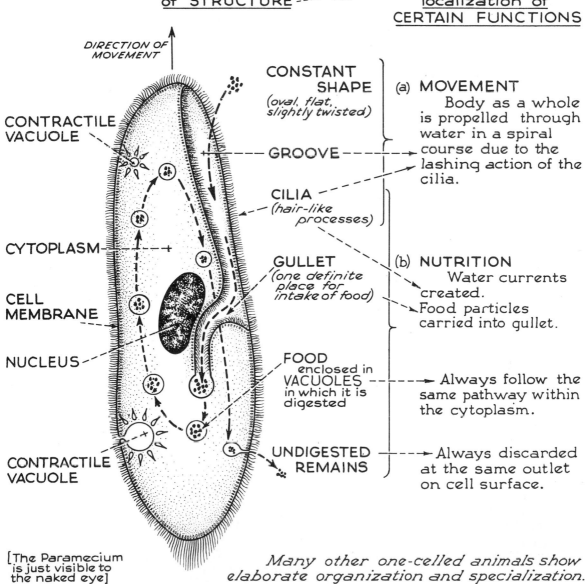

DIRECTION OF MOVEMENT

CONTRACTILE VACUOLE

CYTOPLASM

CELL MEMBRANE

NUCLEUS

CONTRACTILE VACUOLE

CONSTANT SHAPE
(oval, flat, slightly twisted)

GROOVE

CILIA
(hair-like processes)

GULLET
(one definite place for intake of food)

FOOD enclosed in VACUOLES in which it is digested

UNDIGESTED REMAINS

(a) MOVEMENT
   Body as a whole is propelled through water in a spiral course due to the lashing action of the cilia.

(b) NUTRITION
   Water currents created.
Food particles carried into gullet.

Always follow the same pathway within the cytoplasm.

Always discarded at the same outlet on cell surface.

[The Paramecium is just visible to the naked eye]

Many other one-celled animals show elaborate organization and specialization.

The CELL

— STRUCTURAL *and* FUNCTIONAL *unit of the* MANY-CELLED ANIMAL.
   Higher animals, including MAN, are made up of millions of living cells which vary widely in structure and function but have certain features in common.

A GENERALIZED ANIMAL CELL *(of secretory type).* [The finer details are seen only in the high magnification afforded by ELECTRON MICROSCOPY.]

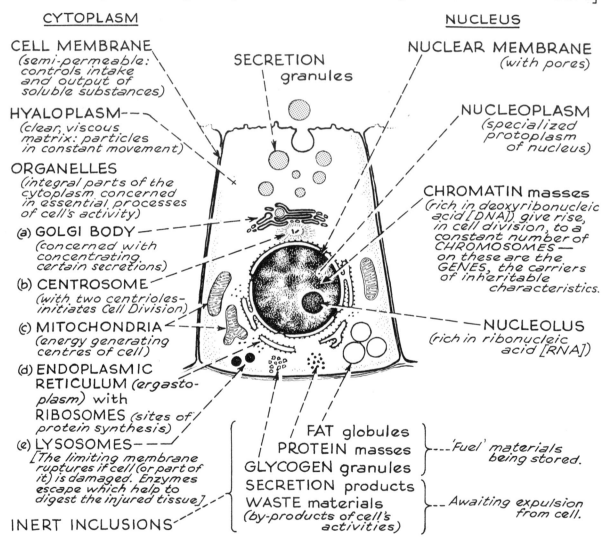

CYTOPLASM

CELL MEMBRANE
(semi-permeable: controls intake and output of soluble substances)

HYALOPLASM—
(clear, viscous matrix: particles in constant movement)

ORGANELLES
(integral parts of the cytoplasm concerned in essential processes of cell's activity)

(a) GOLGI BODY—
(concerned with concentrating certain secretions)

(b) CENTROSOME
(with two centrioles- initiates Cell Division)

(c) MITOCHONDRIA
(energy generating centres of cell)

(d) ENDOPLASMIC RETICULUM *(ergasto- plasm)* with RIBOSOMES *(sites of protein synthesis)*

(e) LYSOSOMES—
[The limiting membrane ruptures if cell (or part of it) is damaged. Enzymes escape which help to digest the injured tissue]

INERT INCLUSIONS

SECRETION
granules

NUCLEUS

NUCLEAR MEMBRANE
(with pores)

NUCLEOPLASM
(specialized protoplasm of nucleus)

CHROMATIN masses
(rich in deoxyribonucleic acid [DNA]) give rise, in cell division, to a constant number of CHROMOSOMES — on these are the GENES, the carriers of inheritable characteristics.

—NUCLEOLUS
(rich in ribonucleic acid [RNA])

FAT globules
PROTEIN masses
GLYCOGEN granules } ----'Fuel' materials being stored.

SECRETION products
WASTE materials
(by-products of cell's activities) } --- Awaiting expulsion from cell.

CELL DIVISION (MITOSIS)

All cells arise from the division of pre-existing cells. In MITOSIS there is an exact QUALITATIVE division of the NUCLEUS and a less exact QUANTITATIVE division of the CYTOPLASM.

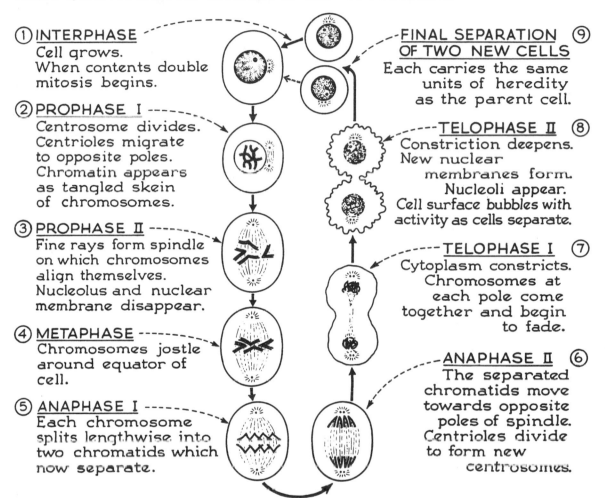

① INTERPHASE
Cell grows.
When contents double mitosis begins.

② PROPHASE I
Centrosome divides.
Centrioles migrate to opposite poles.
Chromatin appears as tangled skein of chromosomes.

③ PROPHASE II
Fine rays form spindle on which chromosomes align themselves.
Nucleolus and nuclear membrane disappear.

④ METAPHASE
Chromosomes jostle around equator of cell.

⑤ ANAPHASE I
Each chromosome splits lengthwise into two chromatids which now separate.

FINAL SEPARATION ⑨ OF TWO NEW CELLS
Each carries the same units of heredity as the parent cell.

TELOPHASE II ⑧
Constriction deepens.
New nuclear membranes form.
Nucleoli appear.
Cell surface bubbles with activity as cells separate.

TELOPHASE I ⑦
Cytoplasm constricts.
Chromosomes at each pole come together and begin to fade.

ANAPHASE II ⑥
The separated chromatids move towards opposite poles of spindle.
Centrioles divide to form new centrosomes.

The longitudinal halving of chromosomes and genes ensures that each new cell receives the same hereditary factors as the original cell.

The number of chromosomes is constant for any one species.
The cells of the human body (somatic cells) carry 23 pairs — i.e. 46 chromosomes.

[For clarity only 4 chromosomes (2 pairs) are shown in these diagrams]

DIFFERENTIATION of ANIMAL CELLS

<u>SPECIALIZATION</u> distinguishes multicellular creatures from more primitive forms of life.

<u>ONE-CELLED ANIMALS</u>   Capable of INDEPENDENT existence —
Undifferentiated — Show all activities or ····· PHENOMENA of LIFE

<u>MANY-CELLED ANIMALS</u>   Cells CO-OPERATE for well-being of whole body.   *All cells retain powers of*
ORGANIZATION
IRRITABILITY
NUTRITION
METABOLISM
RESPIRATION
EXCRETION

Differentiated —— Groups of cells undergo adaptations and sacrifice some powers to fit them for special duties.

<u>MODIFICATION of STRUCTURE</u> ···· *for efficient* ···· <u>SPECIALIZATION of FUNCTION</u> ···· *with* ···· <u>LOSS or REDUCTION of VERSATILITY</u>

E.g.

SECRETORY CELL ········ Highly developed powers of SECRETION e.g. enzymes for chemical breakdown of foodstuffs.   *Diminished powers of* CONTRACTION and REPRODUCTION

Cytoplasm
Nucleus displaced to base by formed and stored secretion
Nucleus

FAT CELL ·············· STORAGE of FAT   *Loss of powers of* CONTRACTION and SECRETION

Cytoplasm
Cytoplasm displaced by stored fat
Nucleus

MUSCLE CELL ············· Highly developed powers of CONTRACTILITY   *Diminished powers of* SECRETION and REPRODUCTION

Cytoplasm   Nucleus
Elongated cell body

NERVE CELL ············· Highly developed powers of IRRITABILITY   *Loss of powers of* REPRODUCTION

Cytoplasm
(response to stimuli and transmission of impulses over long distances)   *i.e. if nerve cell is destroyed no regeneration is possible.*

Cytoplasm drawn out into long branching processes
Nucleus
×500

ORGANIZATION OF TISSUES

Different cell types are not mixed haphazardly in the body.
Cells which are alike are arranged together to form TISSUES.
There are *four* main types of tissue:– 1. EPITHELIA or LINING,
2. CONNECTIVE or SUPPORTING, 3. MUSCULAR, 4. NERVOUS.

EPITHELIA

STRUCTURAL MODIFICATIONS *Sheets of cells with minimum intercellular substance*	SITE *Line all internal and external surfaces of body*	SPECIALIZED FUNCTIONS

A. SIMPLE   Single layer

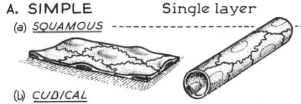

(a) SQUAMOUS - - - - - - - - - - - - - - -Lining of blood vessels, heart and cavities of body.

Reduces friction between surfaces.

(b) CUBICAL

- - - - - - - SIMPLE - - - -Small ducts e.g. of salivary glands.

Protects underlying tissues: non-secretory.

SECRETORY
- - -SEROUS — Many glands e.g. serous secreting part of mixed salivary gland.

Forms watery secretion containing e.g. digestive enzymes.

MUCOUS- -Mucous glands e.g. in mixed salivary gland.

Forms viscous lubricant, protective mucus.

(c) COLUMNAR

- - - - - - - - -Large ducts of kidney.

Protects and lines.

Simple

Mucous

Goblet Cells

- - - - - - - -Surface lining of stomach.

Secretes carpet of mucus to protect stomach from its own acid and digestive enzymes.

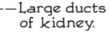

E.g. in lining of intestine.

Form a viscous, lubricating, protective mucus.

Small intestine.

Absorbs foodstuffs.

Striated or Brush Bordered

Ciliated *(Motile hair-like processes at free surface)* Uterine tube.

Forms currents — wafts ovum towards womb.

x500

<u>EPITHELIA</u>

B. PSEUDOSTRATIFIED COLUMNAR CILIATED *(with Goblet cells)* - - - - - - - Respiratory passages.

(a) Protective.

(b) Cilia form currents to move mucus-trapped particles towards the back of the throat.

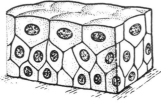

C. COMPOUND - Many layers of cells
TRUE STRATIFIED

(a) *TRANSITIONAL* - Urinary passages.

(a) Extensible.

(b) Protective — prevents penetration of urine into underlying tissues.

(b) *STRATIFIED SQUAMOUS* - - - - - - - - - - - - - Surfaces subjected to great wear and tear.

Most resistant type of epithelium.

(i) *Uncornified*

Internal surfaces e.g. mouth and gullet.

Protects against friction.

(ii) *Cornified* - - - - - - - - - - - External skin surfaces.

Protects against wear and tear, evaporation and extremes of temperature.

Cornified dead cell layers

Clear layer

Granular layer

Prickle cell layers

Germinal layer *(for cell replacement)*

[Dead cell layers vary in thickness — e.g. thickest on soles of feet and palms of hands.]

×400

10

CONNECTIVE TISSUES

STRUCTURAL MODIFICATIONS
Cells plus large amount of intercellular matrix and extracellular elements.

SPECIALIZED FUNCTIONS
Form framework, connecting, supporting and packing tissues of the body.

1. CELLS (floating free) in a FLUID MATRIX

BLOOD

Red Blood Corpuscle *(no nucleus)* ⬤ ⬤ --------------- Oxygen transport to tissues.

White Blood Cells

Granular
- *Neutrophil* ------------ Defence against bacteria.
- *Eosinophil* -------- Limits immune responses.
- *Basophil* ----------- Heparin formation.

Non-granular
- ------- Defence against bacteria.
- *Monocyte*
- *Lymphocytes*

Immunity reactions (Antibody formation).

x 1000

Platelets *(no nuclei)* ------------------ Rôle in blood clotting.

2. CELLS in SEMI-SOLID JELLY-LIKE MATRIX with fine fibrils.

MESENCHYME

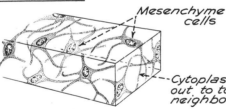

Mesenchyme cells

Cytoplasm drawn out to touch neighbouring cells

Earliest type found in embryo.

From it all other connective tissues differentiate.

3. CELLS in SEMI-SOLID MATRIX with fine network of extracellular RETICULAR FIBRES.

RETICULAR

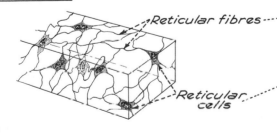

Reticular fibres

Reticular cells

Form 3-dimensional 'net'. Framework of e.g. spleen, lymph nodes, bone marrow.

Potential phagocytes - capable of ingesting particles.

x500

11

CONNECTIVE TISSUES

4. CELLS in SEMI-SOLID MATRIX with thicker collagenous and elastic fibres.

(a) LOOSE FIBROUS

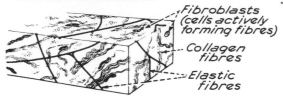

Fibroblasts (cells actively forming fibres)

Collagen fibres

Elastic fibres

"Packing" tissue between organs: sheaths of muscles, nerves and blood vessels.

(b) DENSE FIBROUS

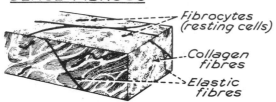

Fibrocytes (resting cells)

Collagen fibres

Elastic fibres

Strong, inelastic yet pliable — e.g. ligaments; capsules of joints; heart valves.

(c) ELASTIC

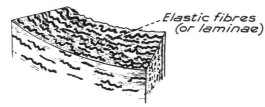

Elastic fibres (or laminae)

Strong, extensible and flexible — e.g. in walls of blood vessels and air passages.

(d) ADIPOSE

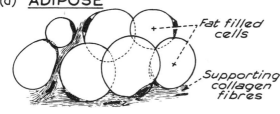

Fat filled cells

Supporting collagen fibres

Protective "cushion" for organs. Insulating layer in skin. Storage of fat reserves.

(e) TENDON

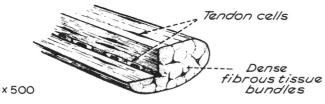

Tendon cells

Dense fibrous tissue bundles

Tough, inelastic cords of dense fibrous tissue which attach muscles to bones.

× 500

12

CONNECTIVE TISSUES

5. CELLS in SOLID ELASTIC MATRIX with fibres.

(a) HYALINE CARTILAGE

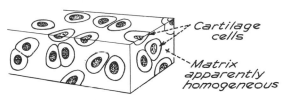

Cartilage cells

Matrix apparently homogeneous

Firm yet resilient
e.g. in air passages; ends of bones at joints.
Transition stage in bone
formation.

(b) WHITE FIBRO-CARTILAGE

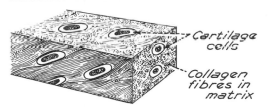

Cartilage cells

Collagen fibres in matrix

Tough. Resistant to stretching.
Acts as shock absorber between vertebrae.

(c) ELASTIC or YELLOW FIBRO-CARTILAGE

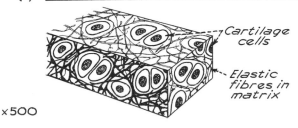

Cartilage cells

Elastic fibres in matrix

More flexible, resilient —
e.g. in larynx and ear.

x500

6. CELLS in SOLID RIGID MATRIX impregnated with Calcium and Magnesium Salts.

BONE

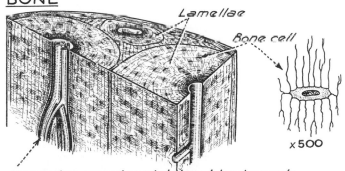

Lamellae

Bone cell

Forms rigid framework
(or skeleton)   of body.

x500

Haversian canal containing blood vessels.

x100

13

MUSCULAR TISSUES

All have ELONGATED cells with special development of CONTRACTILITY.

1. SMOOTH, UNSTRIPED, VISCERAL or INVOLUNTARY muscle

Nucleus

Least specialized. Shows slow rhythmical contraction and relaxation.
Not under voluntary control.
Found in walls of viscera and blood vessels.

2. CARDIAC or HEART muscle

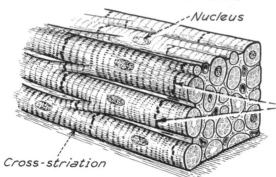

--Nucleus

Cross-striation

More highly specialized. Rapid rhythmical contraction (and relaxation) spreads through whole muscle mass.
Not under voluntary control.
Found only in heart wall.

Cells adhere, end to end, at intercalated discs to form long 'fibres' which branch and connect with adjacent 'fibres'.

(Note: There is no protoplasmic continuity between cells.)

3. STRIATED, STRIPED, SKELETAL or VOLUNTARY muscle

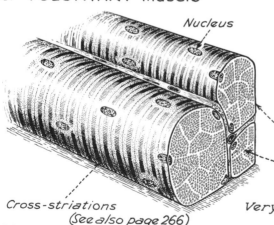

Nucleus

Most highly specialized.
Very rapid, powerful contractions of individual fibres.
Under voluntary control.
Found in e.g. muscles of trunk, limbs, head.

Thick covering membrane (sarcolemma)

Many myofibrils embedded in sarcoplasm

Cross-striations (See also page 266)

x500

Very long, large multi-nucleated units. No branching.

NERVOUS TISSUES

Nervous tissue is divided into:-

(a) <u>NEURONES</u> or *True* nerve cells
    specialized in
     IRRITABILITY
     CONDUCTION
     INTEGRATION

<u>MOTOR</u> — pass messages from
    brain and spinal cord to effector
    organs (muscles and glands)

<u>ASSOCIATION</u> — relay messages
    between neurones.

<u>SENSORY</u> — receive and pass
    messages from environment
    to brain and spinal cord.

(b) <u>ACCESSORY</u> or *Supporting* cells
    Not RECEPTIVE
    Not CONDUCTING

<u>NEUROGLIA</u> in Central Nervous
    System (*Brain and Spinal Cord*)

<u>SHEATH</u> (*Schwann*) cells on peripheral
    nerve fibres, i.e. outside C.N.S.

<u>SATELLITE</u> cells in ganglia of
    peripheral nervous system.

- -

<u>TYPICAL NERVE CELL</u> (*Motor*)

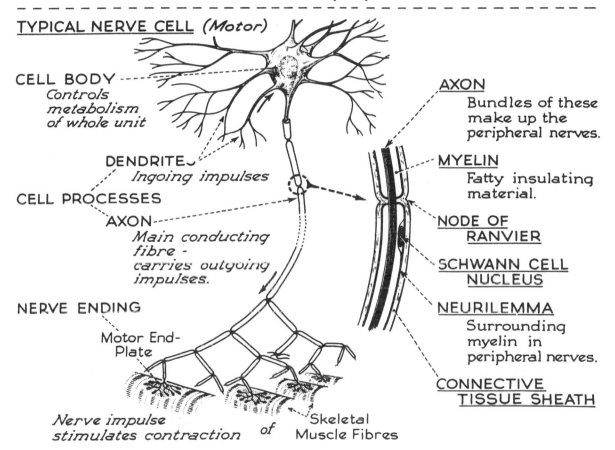

CELL BODY
*Controls
metabolism
of whole unit*

DENDRITE
Ingoing impulses

CELL PROCESSES

AXON
*Main conducting
fibre -
carries outgoing
impulses.*

NERVE ENDING

Motor End-
Plate

*Nerve impulse
stimulates contraction* of
Skeletal
Muscle Fibres

AXON
Bundles of these
make up the
peripheral nerves.

MYELIN
Fatty insulating
material.

**NODE OF
RANVIER**

**SCHWANN CELL
NUCLEUS**

NEURILEMMA
Surrounding
myelin in
peripheral nerves.

**CONNECTIVE
TISSUE SHEATH**

NERVOUS TISSUES

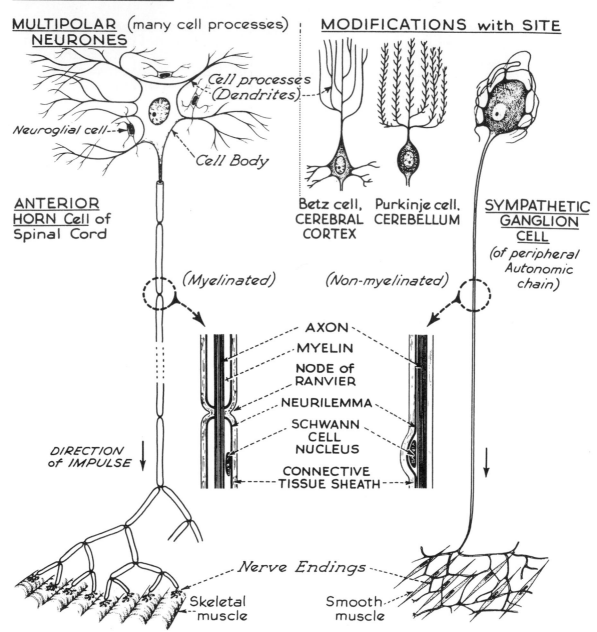

MULTIPOLAR (many cell processes)
NEURONES

Cell processes
(Dendrites)

Neuroglial cell

Cell Body

MODIFICATIONS with SITE

Betz cell,
CEREBRAL
CORTEX

Purkinje cell,
CEREBELLUM

**ANTERIOR
HORN Cell** of
Spinal Cord

(Myelinated)

(Non-myelinated)

**SYMPATHETIC
GANGLION
CELL**
(of peripheral
Autonomic
chain)

AXON
MYELIN
NODE of
RANVIER
NEURILEMMA
SCHWANN
CELL
NUCLEUS
CONNECTIVE
TISSUE SHEATH

*DIRECTION
of IMPULSE*

Nerve Endings

Skeletal
muscle

Smooth
muscle

*Most multipolar neurones are MOTOR (efferent) or
ASSOCIATION in function.*

16

NERVOUS TISSUES

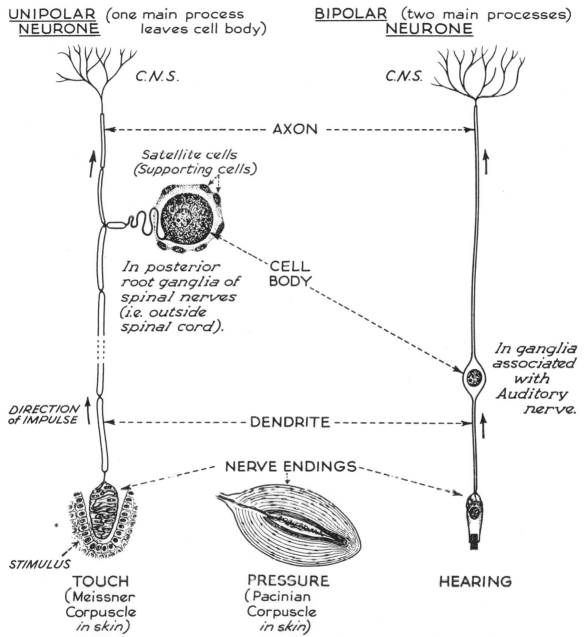

UNIPOLAR NEURONE (one main process leaves cell body)

BIPOLAR NEURONE (two main processes)

C.N.S.

C.N.S.

AXON

Satellite cells (Supporting cells)

In posterior root ganglia of spinal nerves (i.e. outside spinal cord).

CELL BODY

In ganglia associated with Auditory nerve.

DIRECTION of IMPULSE

DENDRITE

NERVE ENDINGS

STIMULUS

TOUCH (Meissner Corpuscle in skin)

PRESSURE (Pacinian Corpuscle in skin)

HEARING

All unipolar and bipolar neurones are SENSORY (afferent) in function.

CELL DIVISION (*MEIOSIS*)

New individuals develop after fusion of 2 specialized cells—the GAMETES. During their formation the OVUM (*female*) and the SPERMATOZOON (*male*) undergo two special cell divisions to reduce the chromosome content of each to the haploid number of 23.   In fusion, the mingling of male and female chromosomes restores the normal diploid number—46.

MATURATION of the MALE GAMETES (*only one pair of chromosomes illustrated*)

FIRST DIVISION

SECOND DIVISION

4 new cells

4 mature sperms

Modified Prophase

Pairing of corresponding chromosomes

2 new cells

Becoming coiled

Metaphase

Anaphase

Telophase

Formation of chromatids

Anaphase

Anaphase

Cross-over and incomplete separation

The chromatids now separate

Metaphase

Chromatids do not separate
Interchange of hereditary material between chromosomes.

[End result:– 4 cells (*all different*) from original germ cell - each with 23 chromosomes]

In the <u>FEMALE</u> , three of the 'cells' are small polar bodies which are discarded and disintegrate.

ONE single mature OVUM receives most of the cytoplasm.

DEVELOPMENT of the INDIVIDUAL

All tissues of the human body are derived from the single cell— the FERTILIZED OVUM.

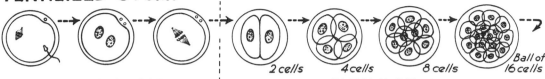

FERTILIZATION
Fusion of ovum and spermatozoon

2 cells   4 cells   8 cells   Ball of 16 cells

CLEAVAGE
Repeated mitotic divisions: each cell receives equal number of *Maternal* and *Paternal* chromosomes.

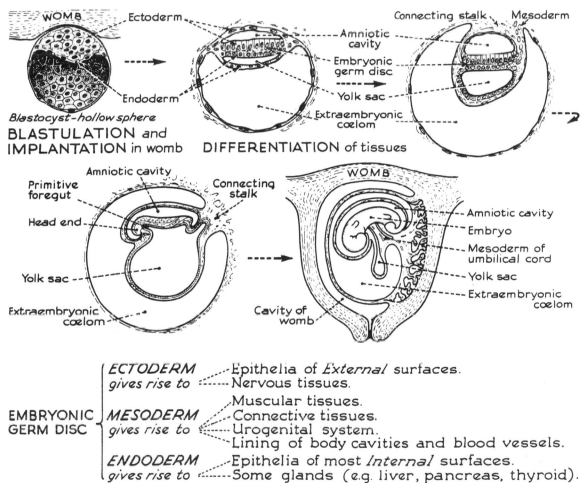

Blastocyst-hollow sphere

BLASTULATION and IMPLANTATION in womb

DIFFERENTIATION of tissues

EMBRYONIC GERM DISC	*ECTODERM* gives rise to	Epithelia of *External* surfaces.
		Nervous tissues.
	MESODERM gives rise to	Muscular tissues.
		Connective tissues.
		Urogenital system.
		Lining of body cavities and blood vessels.
	ENDODERM gives rise to	Epithelia of most *Internal* surfaces.
		Some glands (e.g. liver, pancreas, thyroid).

"MASTER" TISSUES

It is an aid to understanding the general Design of the body to think of the systems as grouped into "MASTER" Tissues and "VEGETATIVE" Systems.

The "MASTER" Tissues are those which specialize in receiving stimuli from the environment and reacting to them —
NERVOUS TISSUE and SKELETAL MUSCLE.

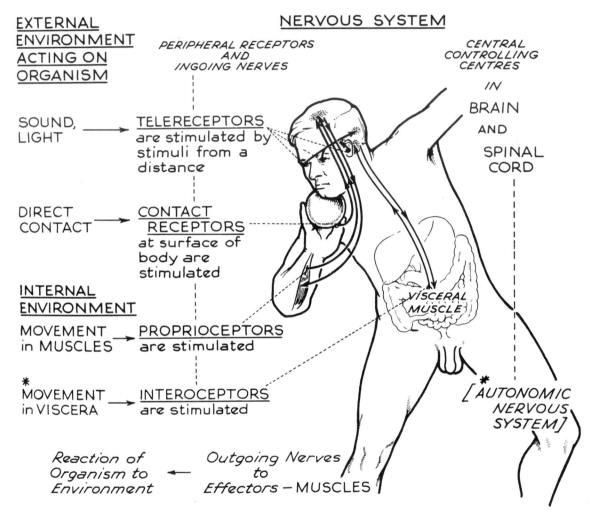

NERVOUS SYSTEM

EXTERNAL ENVIRONMENT ACTING ON ORGANISM

PERIPHERAL RECEPTORS AND INGOING NERVES

CENTRAL CONTROLLING CENTRES

IN

BRAIN

AND

SPINAL CORD

SOUND, LIGHT → TELERECEPTORS are stimulated by stimuli from a distance

DIRECT CONTACT → CONTACT RECEPTORS at surface of body are stimulated

INTERNAL ENVIRONMENT

MOVEMENT in MUSCLES → PROPRIOCEPTORS are stimulated

*MOVEMENT in VISCERA → INTEROCEPTORS are stimulated

VISCERAL MUSCLE

[*AUTONOMIC NERVOUS SYSTEM]

Reaction of Organism to Environment ← Outgoing Nerves to Effectors – MUSCLES

[*Note that a special part of the Nervous System — the AUTONOMIC NERVOUS SYSTEM — regulates the activities of the "Vegetative" Systems.]

"VEGETATIVE" SYSTEMS

The VEGETATIVE Systems are those which serve the basic utilities of life. They deal with the source of energy — FOOD- breaking it down and transporting it around the body to make it available to the cells for Growth and Repair, for Reproduction and for Movement, etc. They also get rid of Waste. Much of their activity goes on below the level of consciousness.

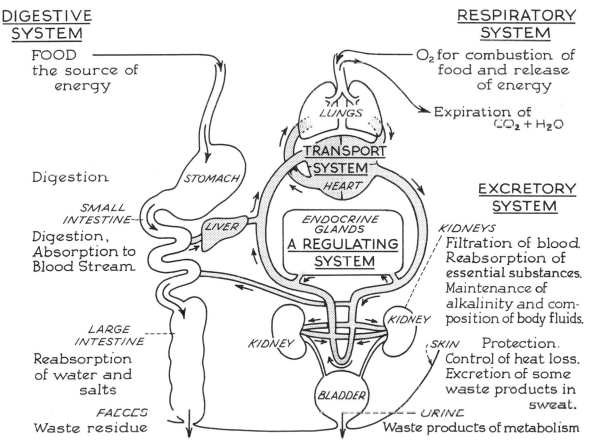

DIGESTIVE SYSTEM

FOOD
the source of
energy

Digestion

SMALL INTESTINE

Digestion,
Absorption to
Blood Stream

LARGE INTESTINE

Reabsorption
of water and
salts

FAECES
Waste residue

RESPIRATORY SYSTEM

O_2 for combustion of
food and release
of energy

Expiration of
$CO_2 + H_2O$

EXCRETORY SYSTEM

KIDNEYS
Filtration of blood.
Reabsorption of
essential substances.
Maintenance of
alkalinity and com-
position of body fluids.

SKIN Protection.
Control of heat loss.
Excretion of some
waste products in
sweat.

URINE
Waste products of metabolism

(Diagram labels: LUNGS, TRANSPORT SYSTEM, HEART, STOMACH, LIVER, ENDOCRINE GLANDS A REGULATING SYSTEM, KIDNEY, KIDNEY, BLADDER)

The activities of these systems are regulated partly through a section of the MASTER tissues - the AUTONOMIC NERVOUS SYSTEM. Its complementary divisions - PARASYMPATHETIC and SYMPATHETIC - carry opposite or antagonistic messages to viscera. These impulses are rapidly adjusted and balanced by the BRAIN CENTRES to integrate the actions of the vegetative systems to the constantly changing needs of the whole body.

The BODY SYSTEMS

The tissues are arranged to form ORGANS.
Organs are grouped into SYSTEMS.

The ESSENTIAL LIFE PROCESSES ----are delegated to-------- SEPARATE SYSTEMS

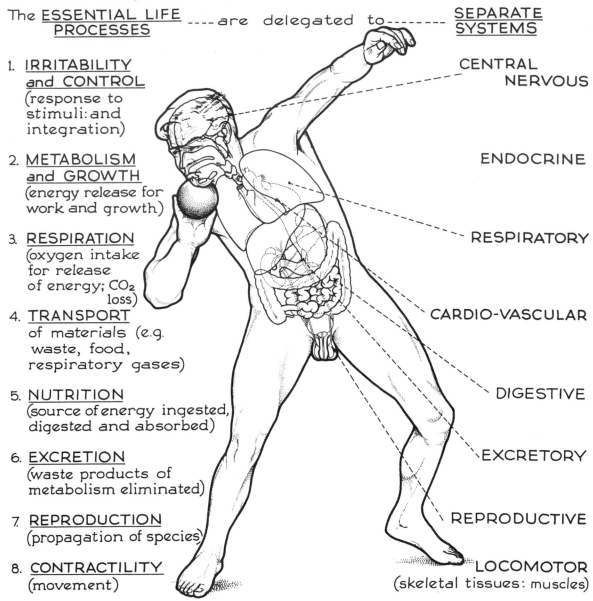

1. IRRITABILITY and CONTROL
(response to stimuli: and integration)

2. METABOLISM and GROWTH
(energy release for work and growth)

3. RESPIRATION
(oxygen intake for release of energy; CO_2 loss)

4. TRANSPORT
of materials (e.g. waste, food, respiratory gases)

5. NUTRITION
(source of energy ingested, digested and absorbed)

6. EXCRETION
(waste products of metabolism eliminated)

7. REPRODUCTION
(propagation of species)

8. CONTRACTILITY
(movement)

CENTRAL NERVOUS

ENDOCRINE

RESPIRATORY

CARDIO-VASCULAR

DIGESTIVE

EXCRETORY

REPRODUCTIVE

LOCOMOTOR
(skeletal tissues: muscles)

*The systems do not work independently. The body works as a whole.
Health and well-being depend on the co-ordinated effort of every part.*

22

CHAPTER 2

NUTRITION and METABOLISM

The SOURCES, RELEASE and
USES of ENERGY

BASIC CONSTITUENTS of PROTOPLASM

Protoplasm is made up of certain
ELEMENTS —— present mainly in —— CHEMICAL COMBINATION

Chemical Symbol

e.g. in a 70 kg man average amounts (grams)

H — HYDROGEN	6,580	
O — OXYGEN	43,550	
C — **CARBON**	12,590	
N — NITROGEN	1,815	

A large amount of the H and O is present as WATER (H_2O)

C, H & O combine chemically to form CARBOHYDRATES and FATS (*chief sources of ENERGY in living protoplasm*)

C, H, O and N combine to form PROTEINS (*main BUILDING constituents of all protoplasm*)

These make up most of the Body Weight

Ca — CALCIUM	1,700	
P — PHOSPHORUS	680	

Important constituents of blood and of hard tissues – *e.g.* bones and teeth.

Cl — CHLORINE	115	
Na — SODIUM	70	

Important constituents of body fluids.

K — POTASSIUM	70	

Important constituent of all cells.

S — SULPHUR	100	
Mg — MAGNESIUM	42	

Important for activity of Brain, Nerves and Muscles.

Fe — IRON	7	

Important component of haemoglobin in red blood cells

These eight make up much of the remaining Body Weight

Trace elements *make up the last few grams or so.*
These include:- Manganese, copper, iodine, zinc, cobalt, molybdenum, nickel, aluminium, chromium, titanium, silicon, rubidium, lithium, arsenic, fluorine, bromine, selenium, boron, barium and strontium.

NOTE:- *Very few of these inorganic substances are found free in protoplasm. Most are in chemical combination. Apart from water, the chief constituents are present as compounds of CARBON, i.e. they are ORGANIC SUBSTANCES.*

CARBOHYDRATES

CARBOHYDRATES are found in Plant and Animal Protoplasm. The simpler carbohydrates are known as SUGARS; the more complex as STARCH. Starch is made up of hundreds of units of sugar tied chemically together. They consist of atoms of C, H and O.

GLUCOSE – $C_6H_{12}O_6$ – is one of the simplest SUGARS —— a MONOSACCHARIDE
[Found in plants and animals]

The arrangement of atoms in the molecule may be represented thus:-

usually written →

LACTOSE – $C_{12}H_{22}O_{11}$ – is another relatively simple SUGAR.
[Milk sugar] Two MONOSACCHARIDE molecules combine to form a –

DISACCHARIDE
with loss of one molecule of WATER (H_2O)

[N.B. Same constituent elements as Glucose but with different spatial arrangement.]

GALACTOSE   GLUCOSE

STARCH – $(C_6H_{10}O_5)_n$ where n > 100 – is built up of long chains of GLUCOSE units to form a-
[Form in which PLANTS store sugar]
POLYSACCHARIDE
with loss of one molecule of WATER at each link

These chains may then be arranged in layers:-

chain   link
chain   link
chain

GLYCOGEN – $C_n(H_2O)_m$ – is built up of 12 to 18 GLUCOSE units similarly linked
[Form in which ANIMALS store sugar] chemically as a POLYSACCHARIDE but with *branching chains*.

FATS

FATS are present in Plant and Animal Tissues. The molecule of fat is made up of smaller molecules of FATTY ACIDS linked chemically with a molecule of GLYCEROL. They consist of atoms of the elements C, H and O.

<u>SIMPLE LIPIDS</u>

<u>GLYCEROL</u> can react with <u>1</u> molecule of a <u>FATTY ACID</u> to produce a <u>MONOGLYCERIDE</u>

e.g. *BUTYRIC ACID* — " ——— *MONOBUTYRIN*

CH_2OH
$CHOH$ + HO $OCCH_2CH_2CH_3$ ⟶ $CH_2OOCCH_2CH_2CH_3$
CH_2O H

with the loss of one molecule of WATER (H_2O)

<u>GLYCEROL</u> can react with <u>2</u> molecules of <u>FATTY ACID</u> to form a ——— <u>DIGLYCERIDE</u>

e.g. *PALMITIC ACID* — " ——— *DIPALMITIN*

CH_2O H HO $OC CH_2CH_2CH_2CH_2CH_2CH_2CH_2CH_2 CH_2 CH_2CH_2CH_2CH_2CH_2CH_3$
$CHOH$ + HO $OC CH_2CH_2CH_2CH_2CH_2CH_2CH_2CH_2 CH_2 CH_2CH_2CH_2CH_2CH_2CH_3$
CH_2O H

with the loss of two molecules of H_2O

In its most common form FAT is made up of one molecule of <u>GLYCEROL</u> linked with <u>3</u> molecules of <u>FATTY ACID</u> to form a ——— <u>TRIGLYCERIDE</u>

e.g. *STEARIC ACID* — " ——— *TRISTEARIN*

CH_2O H HO $OCCH_2CH_2CH_2CH_2CH_2CH_2CH_2CH_2CH_2CH_2CH_2CH_2CH_2CH_2CH_2CH_2CH_3$
CHO H + HO $OCCH_2CH_2CH_2CH_2CH_2CH_2CH_2CH_2CH_2CH_2CH_2CH_2CH_2CH_2CH_2CH_2CH_3$
$CH_2 O$ H HO $OC CH_2CH_2CH_2CH_2CH_2CH_2CH_2CH_2CH_2CH_2CH_2CH_2CH_2CH_2CH_2CH_3$

with the loss of three molecules of H_2O

There are many other naturally occurring FATTY ACIDS. GLYCEROL can combine with 3 molecules of the <u>same</u> or 3 molecules of <u>different</u> FATTY ACIDS to form TRIGLYCERIDES (NEUTRAL FATS).

More complex fat-like substances include

<u>PHOSPHATIDES</u> — *made up of GLYCEROL, FATTY ACIDS, PHOSPHORIC ACID and a NITROGEN base. They are widely distributed in animal tissues.*

<u>CEREBROSIDES</u> — *are even more complex compounds combining SUGAR in their structure.*

COMPOUND LIPIDS

<u>Other fat-like substances</u> *(e.g. Cholesterol, Bile Salts and many Hormones)* are known as STEROIDS

They all share this PARENT STRUCTURE

PROTEINS

PROTEIN is the chief organic constituent of all Protoplasm–Plant or Animal. The protein molecule is made up of smaller units called AMINO-ACIDS. These consist of atoms of C, H, O and N with sometimes S and P.

GLYCINE ———————— is the simplest———————— AMINO-ACID

The arrangement of its atoms may be represented like this:-

[The name 'amino' comes from this NH_2 group] [H—N—C—C—OH (with H below N and C, O double bond)] [The name 'acid' comes from this COOH group]

Both groups are found in all the 20 or so kinds of amino-acid found in nature.

ALANINE ————————is another simple ———————— AMINO-ACID

H—N—C—C—OH
    | |
    H (CH_3) O

All amino-acids share the same core- -N-C-C- . Only the group in this position differs in each.

GLYCINE and
ALANINE can combine easily to form a ——— DIPEPTIDE

(HOH)

H—N—C—C—N—C—C—OH
    | | | |
    H (H) O H (CH_3) O

with the loss of one molecule of WATER (H_2O).

AMINO-ACIDS can link in this way many times to form a POLYPEPTIDE

(H_2O) (H_2O) (H_2O) (H_2O)

H—N—C—C—N—C—C—N—C—C—N—C—C—N—C—C—OH
    | | | | | | | | | |
    H (R) O H (R) O H (R) O H (R) O H (R) O

each link causing loss of one molecule of WATER (H_2O)

Any of the 20 Amino-acids may be linked like this. Long chains with 50 to 400 molecules of amino-acids may make up a molecule of PROTEIN. Note:- The -N-C-C- cores of the amino acids form the backbone of the protein chain.

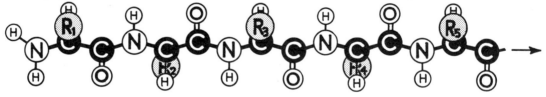

There are thousands of different kinds of Protein but only about 20 different amino-acids. Differences between proteins depend on the amino-acids present and on their number and arrangement.

The various amino-acids can be present in any proportion, arranged in any order, so that the resulting PROTEIN has its own sequence of groups Ⓡ branching off along the sides — this sequence gives it its own special properties which distinguish it from other proteins. The number of possible variations accounts for the great variety of living species, each with a protoplasm specific to itself.

NUCLEIC ACIDS

NUCLEOPROTEINS *(Protein attached to NUCLEIC ACID)* are amongst the most important constituents of Cells and their Nuclei.

NUCLEIC ACIDS are important in controlling heredity. They have a complex chain-like structure built up from MONONUCLEOTIDES —
e.g. Deoxyadenylic acid.

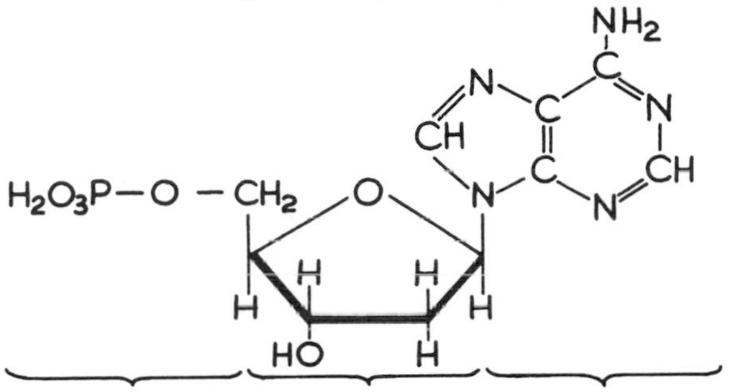

A
NUCLEOTIDE
contains

$H_2O_3P-O-CH_2$

MOLECULES of:—

Phosphoric —— *Sugar* —— *Purine base*
acid (Deoxyribose- [*Adenine (A)*]
a sugar with [*Guanine (G)*]
5 C atoms) —*Pyrimidine base*
[*Cytosine*]
[*Thymine (T)*]

These mononucleotides may condense together to form <u>POLYNUCLEOTIDES</u>.

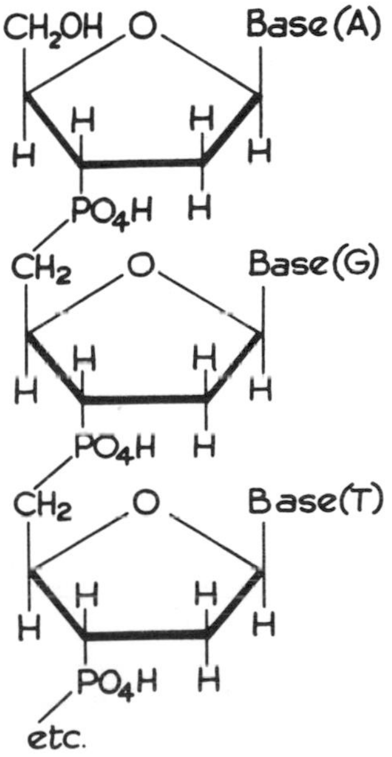

CH_2OH O Base (A)

PO_4H

CH_2 O Base (G)

PO_4H

CH_2 O Base (T)

PO_4H

etc.

Such a chain of nucleotides *(in the DNA molecule in Chromosomes)*, by the specific arrangement of its bases, carries information from one generation of cells to the next.

[The information thus carried in a polynucleotide chain may determine many genetic factors, e.g. whether a child will have brown or blue eyes.]

29

SOURCE of ENERGY: PHOTOSYNTHESIS

    The essential life processes or the phenomena which characterize life depend on the use of ENERGY. The SUN is the SOURCE of ENERGY for ALL LIVING THINGS.

    Only GREEN PLANTS can TRAP and STORE the sun's energy and build from relatively simple substances the energy-rich and body-building compounds — *CARBOHYDRATES, FATS and PROTEINS* — required by PROTOPLASM. This process is called PHOTOSYNTHESIS.

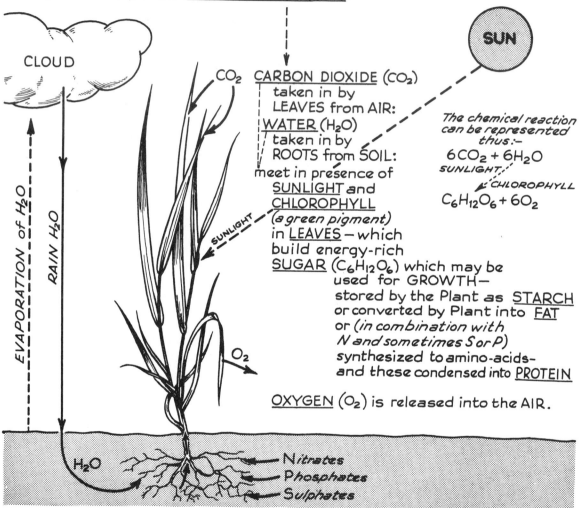

CLOUD

CO₂

CARBON DIOXIDE (CO₂)
  taken in by
LEAVES from AIR:
WATER (H₂O)
  taken in by
ROOTS from SOIL:
meet in presence of
SUNLIGHT and
CHLOROPHYLL
(a green pigment)
in LEAVES — which
build energy-rich
SUGAR (C₆H₁₂O₆) which may be
    used for GROWTH —
    stored by the Plant as STARCH
    or converted by Plant into FAT
    or *(in combination with
    N and sometimes S or P)*
    synthesized to amino-acids —
    and these condensed into PROTEIN

OXYGEN (O₂) is released into the AIR.

The chemical reaction can be represented thus:-

$$6CO_2 + 6H_2O$$

SUNLIGHT   *CHLOROPHYLL*

$$C_6H_{12}O_6 + 6O_2$$

EVAPORATION of H₂O

RAIN H₂O

SUNLIGHT

O₂

H₂O

Nitrates
Phosphates
Sulphates

    When Plants or their products are eaten this stored energy becomes available to animals and man.

CARBON 'CYCLE' in NATURE

Animal bodies are unable to build *Proteins, Carbohydrates* or *Fats* from the raw materials. They must be built up for them by *Plants*.
CARBON is the basic building unit of all these compounds.

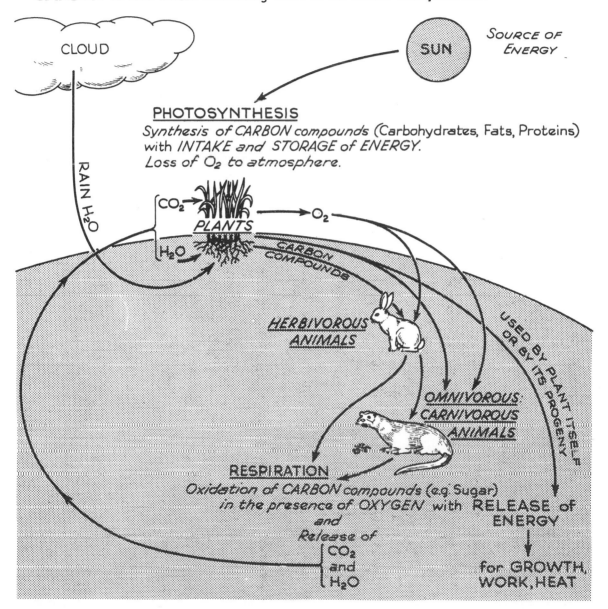

CLOUD

SUN   *SOURCE OF ENERGY*

RAIN H_2O

PHOTOSYNTHESIS
Synthesis of CARBON compounds (Carbohydrates, Fats, Proteins) with INTAKE and STORAGE of ENERGY.
Loss of O_2 to atmosphere.

CO_2
PLANTS
O_2
H_2O
CARBON COMPOUNDS

HERBIVOROUS ANIMALS

OMNIVOROUS CARNIVOROUS ANIMALS

USED BY PLANT ITSELF OR BY ITS PROGENY

RESPIRATION
Oxidation of CARBON compounds (eg Sugar) in the presence of OXYGEN with RELEASE of ENERGY
and
Release of
CO_2
and
H_2O

for GROWTH, WORK, HEAT

NITROGEN 'CYCLE' in NATURE

Although animals are surrounded by NITROGEN in the air they can only use the NITROGEN trapped by Plants.

NITROGEN in the Atmosphere

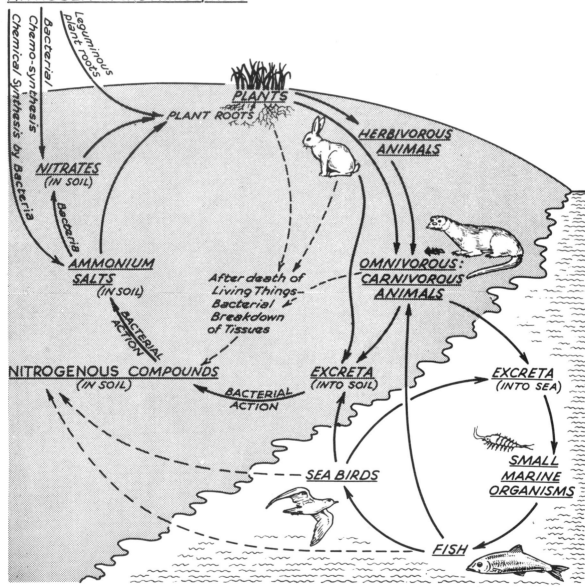

NUTRITION

Man eats as <u>FOOD</u> the substances which PLANTS
(and, through them, ANIMALS) have made.
These are broken down in man's body
into simpler chemical units which provide

BUILDING and PROTECTIVE MATERIALS

Man requires more or less the same
elements as plants. (*Some, such as
the minerals iodine, sodium, iron, calcium,
he assimilates in INORGANIC FORM.*)

Most must be built up for him
by plants:-
ORGANICALLY COMBINED
Carbon, Nitrogen, and Sulphur, etc.
ESSENTIAL AMINO ACIDS

ESSENTIAL FATTY ACIDS
Certain VITAMINS
These are
<u>USED to</u>
<u>BUILD, MAINTAIN or REPAIR</u>
PROTOPLASM

- - - - - - - - - - - - - - - - - -

BODY BUILDING REQUIREMENTS
of the INDIVIDUAL determine
<u>QUALITY</u>
of DIET

ENERGY

*Stored originally
by plants*   <u>RELEASED</u> in man's cells
by
<u>OXIDATION</u>

*When food is 'burned' it gives up
its stored energy*

PROTEINS	yield 4·1 kilocalories	*Units of Energy per gram*
CARBOHYDRATES	yield 4·1 kilocalories	
FATS	yield 9·2 kilocalories	

Most of this appears as <u>HEAT</u>
and is <u>USED for</u>
<u>KEEPING BODY WARM;</u>
some is used for <u>WORK</u> of Cells

- - - - - - - - - - - - - - - - - -

ENERGY REQUIREMENTS
of the INDIVIDUAL determine
<u>QUANTITY</u>
of DIET

For a
<u>BALANCED DIET</u>
TOTAL INTAKE
of
Essential Constituents and Energy Units
must balance...
...Amounts *stored* plus amounts *lost* from body plus amounts *used*
as Work or Heat.

ENERGY-GIVING FOODS

All the main foodstuffs yield ENERGY – the energy originally trapped by Plants. CARBOHYDRATES and FATS are the chief energy-giving foods. PROTEINS can give energy but are mainly used for building and repairing protoplasm.

CARBOHYDRATES are the
PRIMARY SOURCE of ENERGY –
More easily and quickly digested and utilized than Fats

FATS are a
SECONDARY SOURCE of ENERGY –
Ideal energy storage material. Weight for weight, give twice as many energy units as Carbohydrates.

PLANT SOURCES:
SUGAR is found in
LEAVES, FRUITS and ROOTS of PLANTS
and in
FOODS made from them by man
e.g. JAM, TREACLE, SWEETS, SYRUP

i.e. especially those products made by plants for development of next generation

SUGAR can be stored in plants as
CARBOHYDRATE or converted to FAT and stored as such

STARCH is found in
GRAIN, SEEDS and ROOTS of PLANTS
e.g. WHEAT   e.g. POTATOES

OILS in
SEEDS
e.g. OLIVES, COTTON SEEDS

NUTS
e.g. PEANUTS, COCONUTS

and in
FOODS made from them by man
FLOUR e.g. BREAD, CAKES.
CEREALS e.g. CORNFLAKES

COOKING FATS   PEANUT BUTTER

ANIMAL SOURCES:
SUGAR (glucose) is found in TISSUES and BLOOD of ANIMALS and in
FOODS made from them by man
e.g. CHOP   e.g. BLACK PUDDING

and in *Products made by Animals for themselves or the next generation.*

MILK SUGAR (lactose) --------   MILK FAT

HONEY (fructose)   MILK ----------→ BUTTER, CHEESE

SUGAR can be stored in animals as
CARBOHYDRATE or converted to FAT and stored (together with FAT built from Dietary Fatty Acids and Glycerol) in FAT DEPOTS of body.

ANIMAL STARCH (glycogen)
is found in *MUSCLES:*   *LIVER:*
e.g. STEAK   e.g. SUET, LARD, MUTTON FAT, FISH OIL

and in *FOODS made from them by man*
e.g. SAUSAGES   e.g. MARGARINE

BODY-BUILDING FOODS

PROTEIN is the chief body-building food. Because it is the chief constituent of protoplasm, tissues of Plants and Animals are the richest sources.

PLANT SOURCES

Protoplasm of Plant Tissues and — *Stores made by Plants* — or *Foods made*
for the next generation *from them by man.*
in seeds and roots

LEAVES

"2ND CLASS PROTEINS"
–a good source of
AMINO-ACIDS – but
not yielding the full
range essential for
man's growth

e.g. CABBAGE

e.g. PEAS,   *POTATOES*
BEANS

WHEAT → FLOUR → BREAD
e.g.

ANIMAL SOURCES

Protoplasm of Animal Tissues and — *Products made by Animals* — or *Foods made*
to provide for the next *from them by man.*
generation

BEEF

"1ST CLASS PROTEINS"
–built up from a
range of AMINO-
ACIDS similar to
those needed for
GROWTH or REPAIR
of man's own tissues.

POULTRY

MILK

e.g. CHEESE

EGGS

e.g. EGG CUSTARD

FISH

ROE

e.g. MARGARINES
from some fish oils

These foods between them also contain other important body-building elements:- e.g. CALCIUM and PHOSPHORUS for making BONES and TEETH hard. IRON for building HAEMOGLOBIN– the OXYGEN CARRYING substance in RED BLOOD CELLS. IODINE for building the THYROID HORMONE.

PROTECTIVE FOODS — I

Even when provided with BODY-BUILDING and ENERGY-GIVING FOODS, human tissues cannot build or repair their protoplasm or release the energy in the absence of PROTECTIVE substances called <u>VITAMINS</u>. These appear to be important links in METABOLIC processes. Vitamins cannot be substituted for one another in the way that different Carbohydrates, Proteins or Fats can replace others.

ALL are essential for NORMAL GROWTH, HEALTH and WELL-BEING and for general RESISTANCE to INFECTION.

EACH has additional specific protective functions:—

<u>FAT-SOLUBLE VITAMINS</u>	**A** ANTIXEROPH-THALMIC ($C_{20}H_{30}O$)	**D** ANTI-RACHITIC ($C_{28}H_{44}O$)	**E** ANTISTERILITY (Tocopherols $\alpha\beta\gamma$)	**K** ANTI-HAEMORRHAGIC
	Protects SKIN and MUCOUS MEMBRANES (especially front of eye and lining of Digestive and Respiratory tracts). Essential for Regeneration of VISUAL PURPLE.	Important in Ca and P Metabolism. Essential for deposition Ca and P in Bones and Teeth. Promotes absorption of Ca from intestine.	Perhaps important for normal Reproduction in adult. Influences Biochemical Processes in Nuclei of growing cells.	Essential for production of Prothrombin and Factor VII in Liver — important for normal blood clotting.
<u>PLANT SOURCES</u> LEAVES — SEEDS & FRUITS — ROOTS —	Present as precursor CAROTENE. Green, Yellow Vegetables (esp. SPINACH and KALE). Yellow Maize, Peas, Beans. <u>CARROTS</u>.	Vegetables } contain fruits and } negligible cereals } amounts.	Green leaves (e.g. LETTUCE) PEAS. Richest source — GERM of various Cereals e.g. <u>WHEAT GERM OIL</u>.	Spinach, Kale, Cabbage, Cauliflower, Cereals, Tomatoes, Carrots, Potatoes.
<u>ANIMAL SOURCES</u> TISSUES — PRODUCTS for PROGENY FOODS made from them	Animal liver breaks down CAROTENE ($C_{40}H_{56} + 2H_2O \rightarrow 2C_{20}H_{30}O$) Stored in LIVER of Animals and <u>FISH</u> Secreted in MILK EGG YOLK BUTTER CREAM	Formed when ultra-violet rays fall on Sterols in Man's Skin. Stored in LIVER of <u>FISH</u> and Animals MILK BUTTER CREAM CHEESE EGGS	Small amounts in MEAT and DAIRY PRODUCE Vitamin content of dairy produce varies with plant in animal's diet	Synthesized by BACTERIA in man's INTESTINE then absorbed in presence of BILE.

COD LIVER OIL

<u>DIETARY DEFICIENCY</u>	**XEROPHTHALMIA**	**RICKETS** in CHILD	Little evidence of DEFICIENCY syndrome in human beings. (In some female animals developing embryos die, and, in male, testes may show degenerative changes.)	If DEFICIENT absorption from intestine ⟶ upset in mechanism for Blood Clotting
of any or all vitamins retards GROWTH and leads to SEVERE DISEASE and eventually DEATH	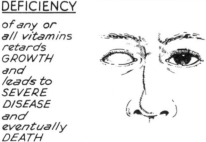	 <u>OSTEOMALACIA</u> in ADULT		

PROTECTIVE FOODS — II

WATER-SOLUBLE VITAMINS

B Complex

Together form essential parts of tissue ENZYME SYSTEMS—concerned in metabolism and Release of Energy by the cells

Includes
B₁ (Aneurin Thiamine) ANTINEURITIC
B₂ RIBOFLAVIN
NICOTINIC ACID
} Each forms part of a COENZYME SYSTEM —concerned with RELEASE of ENERGY from FOOD-STUFFS.

B₆ PYRIDOXINE- Coenzyme in metabolism of certain amino acids.

PANTOTHENIC ACID
BIOTIN — Tissue enzyme in Carbohydrate metabolism.
CHOLINE } Lipotrophic factors-probably
INOSITOL } promote utilization of fats.
PARA-AMINOBENZOIC ACID – No known rôle in man.
FOLIC ACID- An **ANTIANAEMIA** factor.
B₁₂ COBALAMINE-**ANTIPERNICIOUS ANAEMIA** factor—Involved in maturation of red blood cells

C **ANTISCORBUTIC** ($C_6H_8O_6$)

Probably acts as a Respiratory Catalyst in cells. Essential for formation and maintenance of INTERCELLULAR CEMENT and CONNECTIVE TISSUE. (Especially necessary for healthy state of blood vessels)

PLANT SOURCES

Common natural sources for most members of B Complex:-
PULSES e.g. GREEN PEAS (B₁, B₂)
SEEDS and OUTER COATS OF GRAIN e.g. rice, wheat (B₁, B₂, Nic.A)
NUTS e.g. PEANUTS (B₁)
YEAST and Yeast Extracts (B₁, B₂, Nicotinic Acid)

Green vegetables, e.g. Parsley. Citrus fruits, e.g. oranges, lemons, grapefruits, limes. Tomatoes, Rosehips, Black-currants, Red & Green Peppers. Peas. Turnips, Potatoes.

ANIMAL SOURCES

B₂ and Nicotinic Acid are Synthesized by BACTERIA in the human intestine. Found in LIVER (B₁, B₂, Nicotinic Acid, B₁₂, etc.), BACON and LEAN MEAT, MEAT Extract, MILK (B₂, Nicotinic Acid) EGGS (B₂) CHEESE (B₂)

Stored in body — High concentration in Adrenal Glands. Found in Meat, Liver. Secreted in MILK.

DIETARY DEFICIENCY

Deficiency in Pyridoxine, Pantothenic Acid, Biotin Choline Para-amino benzoic A. not observed in man.

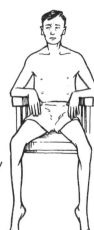

B₁ BERI-BERI
Gross disturbance of function of NERVOUS TISSUE with great MUSCULAR PARALYSIS and WEAKNESS and disturbance in SENSATION. OEDEMA. Gastro-intestinal upsets. HEART FAILURE

B₂ RIBOFLAVINOSIS
Roughening of skin. Cornea becomes cloudy. Cracks and fissures around lips & tongue.

NICOTINIC ACID PELLAGRA
Roughening and reddening of skin. Tongue red and sore. In severe cases-gastro-intestinal upsets and mental derangement

SCURVY
Intercellular cement breaks down. Capillary walls leak→ haemorrhages into tissues - e.g. gums swell and bleed easily. Wounds heal slowly.

B₁₂ PERNICIOUS ANAEMIA
Members of B Complex frequently absent together giving grave disturbances of chemical reactions in all tissues.

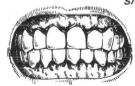

DIGESTION

The organic substances of man's food are chemically <u>similar</u> to those which his body will form from them.   In detail they <u>differ</u>.

Conversion of <u>FOOD SUBSTANCE</u> ——— to ——— <u>BODY SUBSTANCE</u>

Chemical BREAKDOWN
to simpler organic units
in DIGESTIVE TRACT under the action of
ENZYMES

PEPSIN

TRYPSIN
CHYMOTRYPSIN

e.g.
PROTEIN: *EGG* —*breakdown*→ AMINO-ACIDS – – –→*Absorption to blood stream.*
contains protein-   *Synthesis by Body's Cells to*
<u>ALBUMIN</u>   *e.g.* <u>SERUM ALBUMIN</u>
                                                                                            in liver.
               – – – Both Proteins built up – – –
               from about 20 Amino-Acids

AMYLASE { Saliva in Mouth
          Pancreatic Juice
          (Intestinal Juice)

MALTASE

CARBOHYDRATE: *POTATO* —*breakdown*→ GLUCOSE – – – – →*Absorption: Synthesis by*
                 contains   (MONOSACCHARIDE)   *muscle and liver cells to*
                 <u>STARCH</u>   <u>GLYCOGEN</u>
                 – – Both Polysaccharides built – – –
                 up from Glucose units

LIPASE

FAT: <u>*BUTTER*</u> —*Partial breakdown*→ GLYCERIDES – – –→*Absorption to lymph or blood*
                  or   *stream: Synthesis to*
            —*Total breakdown*– –→ FATTY ACIDS   <u>BODY FAT</u>
                         and GLYCEROL
            Both Triglycerides built up
            from Fatty Acids and Glycerol

DIGESTION *is brought about by* <u>SPECIFIC ENZYMES</u> *themselves made of Protein — each acts as a CATALYST for speeding up one particular chemical breakdown without effect on any others.*

PROTEIN METABOLISM

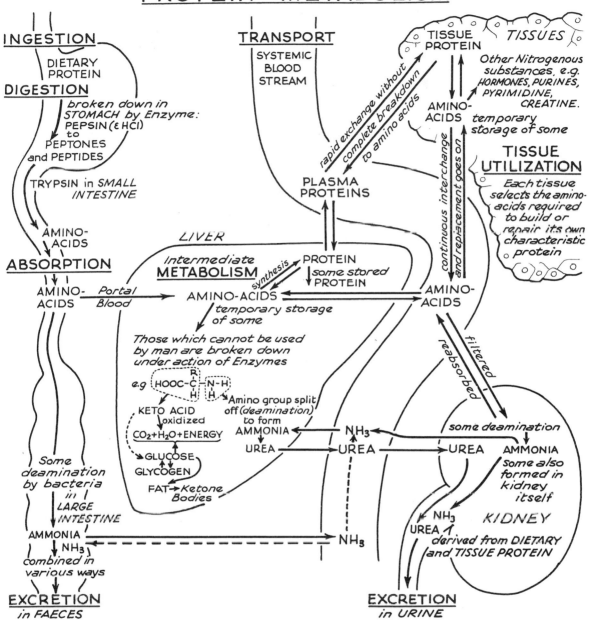

PROTEIN METABOLISM *is under control of Hormones – chiefly the* GROWTH HORMONE (*SOMATOTROPHIN*) *of the* ANTERIOR PITUITARY.

CARBOHYDRATE METABOLISM

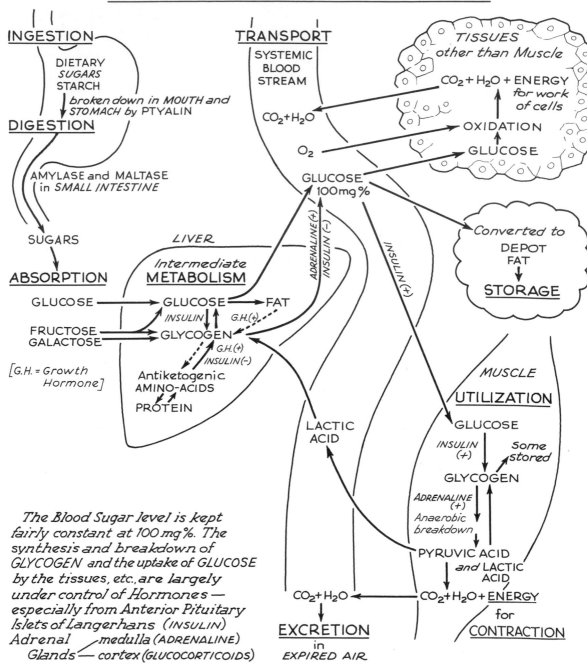

INGESTION

DIETARY
SUGARS
STARCH

broken down in MOUTH and
STOMACH by PTYALIN

DIGESTION

AMYLASE and MALTASE
in *SMALL INTESTINE*

SUGARS

ABSORPTION

GLUCOSE

FRUCTOSE
GALACTOSE

[*G.H. = Growth*
Hormone]

LIVER

Intermediate
METABOLISM

GLUCOSE → FAT

INSULIN   *G.H.(+)*

GLYCOGEN

G.H.(+)
INSULIN(−)

Antiketogenic
AMINO-ACIDS

PROTEIN

TRANSPORT

SYSTEMIC
BLOOD
STREAM

$CO_2 + H_2O$

O_2

GLUCOSE
100mg %

ADRENALINE(+)
INSULIN(−)

INSULIN

INSULIN(+)

LACTIC
ACID

TISSUES
other than Muscle

$CO_2 + H_2O + ENERGY$
for work
of cells

OXIDATION

GLUCOSE

Converted to
DEPOT
FAT

STORAGE

MUSCLE

UTILIZATION

GLUCOSE

INSULIN
(+)

Some
stored

GLYCOGEN

ADRENALINE
(+)
Anaerobic
breakdown

PYRUVIC ACID
and LACTIC
ACID

$CO_2 + H_2O \leftarrow$ $CO_2 + H_2O + ENERGY$
for
EXCRETION   CONTRACTION
in
EXPIRED AIR

The Blood Sugar level is kept
fairly constant at 100 mg %. The
synthesis and breakdown of
GLYCOGEN and the uptake of GLUCOSE
by the tissues, etc., are largely
under control of Hormones —
especially from Anterior Pituitary
Islets of Langerhans (INSULIN)
Adrenal ⎧ medulla (ADRENALINE)
Glands ⎩ cortex (GLUCOCORTICOIDS)

40

FAT METABOLISM

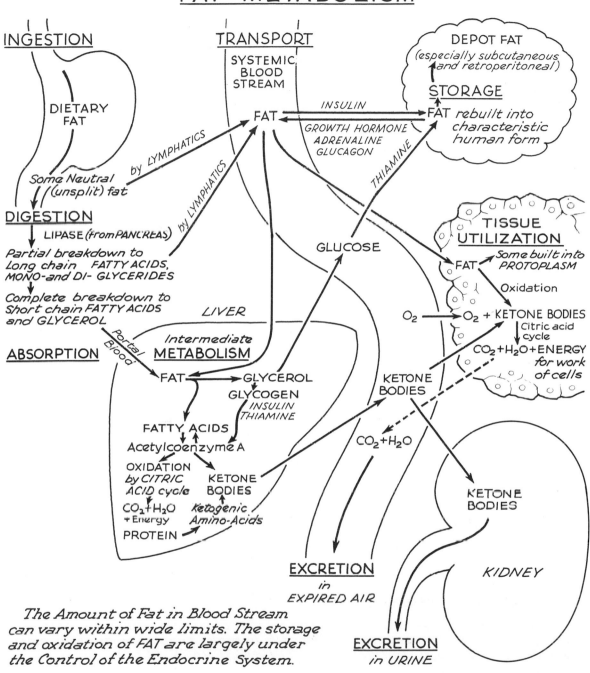

INGESTION

DIETARY FAT

Some Neutral (unsplit) fat

by LYMPHATICS

DIGESTION
LIPASE (from PANCREAS)

Partial breakdown to
Long chain FATTY ACIDS,
MONO- and DI- GLYCERIDES

Complete breakdown to
Short chain FATTY ACIDS
and GLYCEROL

ABSORPTION

Portal Blood

LIVER
Intermediate METABOLISM

FAT → GLYCEROL
GLYCOGEN
INSULIN
THIAMINE

FATTY ACIDS
Acetylcoenzyme A

OXIDATION
by CITRIC
ACID cycle

CO_2+H_2O
+Energy

PROTEIN

KETONE BODIES
Ketogenic
Amino-Acids

TRANSPORT
SYSTEMIC BLOOD STREAM

FAT

INSULIN

GROWTH HORMONE
ADRENALINE
GLUCAGON

by LYMPHATICS

GLUCOSE

KETONE BODIES

CO_2+H_2O

DEPOT FAT
(especially subcutaneous
and retroperitoneal)

STORAGE
FAT rebuilt into
characteristic
human form

THIAMINE

TISSUE UTILIZATION
FAT → Some built into PROTOPLASM

Oxidation

O_2 → O_2 + KETONE BODIES
Citric acid cycle
$CO_2+H_2O+ENERGY$
for work
of cells

KETONE BODIES

KIDNEY

EXCRETION
in
EXPIRED AIR

EXCRETION
in URINE

The Amount of Fat in Blood Stream
can vary within wide limits. The storage
and oxidation of FAT are largely under
the Control of the Endocrine System.

41

RELEASE of ENERGY

ENERGY is released from FOODSTUFFS by OXIDATION.

ENZYME SYSTEMS exist within cells which can convert (in a series of steps) Fats, Proteins or Carbohydrates into compounds suitable for entering the "ENERGY-PRODUCING CYCLE".

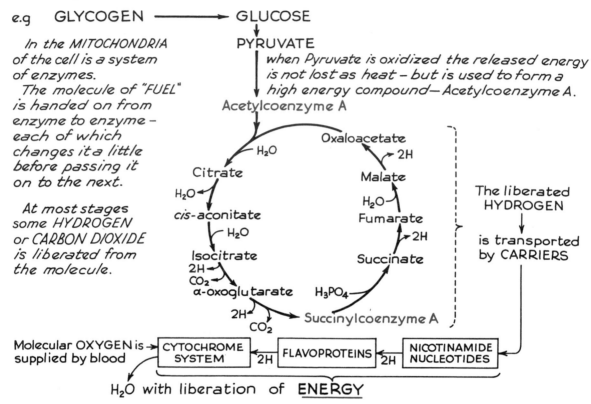

e.g GLYCOGEN ⟶ GLUCOSE

In the MITOCHONDRIA of the cell is a system of enzymes.

The molecule of "FUEL" is handed on from enzyme to enzyme – each of which changes it a little before passing it on to the next.

At most stages some HYDROGEN or CARBON DIOXIDE is liberated from the molecule.

PYRUVATE

when Pyruvate is oxidized the released energy is not lost as heat – but is used to form a high energy compound—Acetylcoenzyme A.

Acetylcoenzyme A

H_2O
Citrate
H_2O
cis-aconitate
H_2O
Isocitrate
2H
CO_2
α-oxoglutarate
2H
CO_2
Succinylcoenzyme A
H_3PO_4
Succinate
2H
Fumarate
H_2O
Malate
2H
Oxaloacetate

The liberated HYDROGEN

is transported by CARRIERS

Molecular OXYGEN is supplied by blood

| CYTOCHROME SYSTEM | 2H | FLAVOPROTEINS | 2H | NICOTINAMIDE NUCLEOTIDES |

H_2O with liberation of **ENERGY**

In the body this ENERGY, instead of being lost as heat, is used to link an additional phosphate group to an existing molecule of ADP *(adenosine diphosphate)* to form the energy-rich compound ATP These compounds are stored and the energy 'trapped' in them is used as required —
e.g.
to *build* more complex STORAGE, STRUCTURAL or FUNCTIONAL materials such as Protein of new tissue in growth and repair, Secretions, Enzymes, etc.
for *work of the cell* – in ways not fully understood – e.g. for muscle contraction and transmission of nerve impulse.
to *liberate heat* to keep body warm.

This KREBS CITRIC-ACID CYCLE is the chief pathway by which potential energy stored in food is released by cells to make their own energy-rich Phosphorus compound–ADENOSINE TRIPHOSPHATE (ATP).

HEAT BALANCE

ENERGY is released in cells by OXIDATION. It is used for work and to keep body warm. The Body Temperature is kept relatively constant *(with a slight fluctuation throughout the 24 hours)* in spite of wide variations in Environmental Temperature and Heat Production.

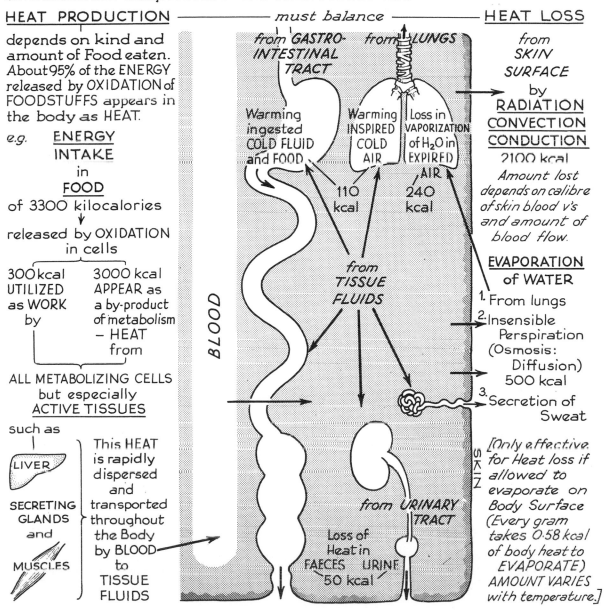

HEAT PRODUCTION ——— *must balance* ——— HEAT LOSS

HEAT PRODUCTION

depends on kind and amount of Food eaten. About 95% of the ENERGY released by OXIDATION of FOODSTUFFS appears in the body as HEAT.

e.g. **ENERGY INTAKE** in **FOOD** of 3300 kilocalories

↓

released by OXIDATION in cells

| 300 kcal UTILIZED as WORK by | 3000 kcal APPEAR as a by-product of metabolism – HEAT from |

ALL METABOLIZING CELLS but especially ACTIVE TISSUES

such as

LIVER

SECRETING GLANDS and

MUSCLES

This HEAT is rapidly dispersed and transported throughout the Body by BLOOD to TISSUE FLUIDS

from GASTRO-INTESTINAL TRACT

from LUNGS

Warming ingested COLD FLUID and FOOD

Warming INSPIRED COLD AIR

Loss in VAPORIZATION of H_2O in EXPIRED AIR

110 kcal

240 kcal

from TISSUE FLUIDS

BLOOD

SKIN

from URINARY TRACT

Loss of Heat in FAECES URINE 50 kcal

HEAT LOSS

from SKIN SURFACE by **RADIATION CONVECTION CONDUCTION** 2100 kcal

Amount lost depends on calibre of skin blood v's and amount of blood flow.

EVAPORATION of WATER

1. From lungs
2. Insensible Perspiration (Osmosis: Diffusion) 500 kcal
3. Secretion of Sweat

[*Only effective for Heat loss if allowed to evaporate on Body Surface* (Every gram takes 0.58 kcal of body heat to EVAPORATE) *AMOUNT VARIES with temperature.*]

43

MAINTENANCE of BODY TEMPERATURE

Any tendency for the BODY TEMPERATURE to *RISE*
as by

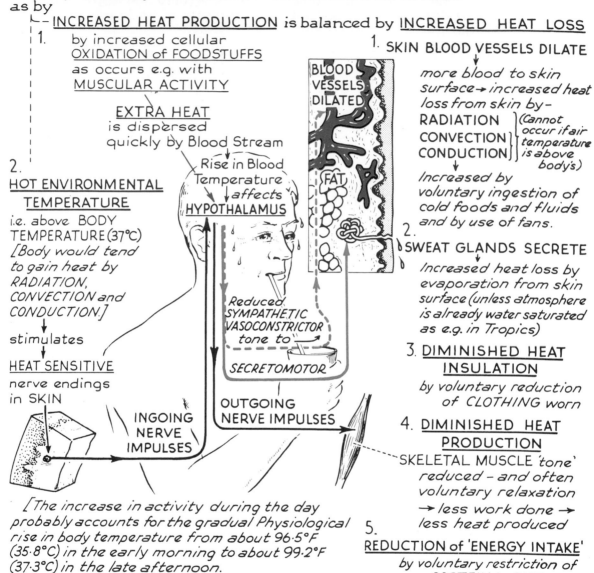

<u>INCREASED HEAT PRODUCTION</u> is balanced by <u>INCREASED HEAT LOSS</u>

1. by increased cellular
<u>OXIDATION of FOODSTUFFS</u>
as occurs e.g. with
<u>MUSCULAR ACTIVITY</u>

<u>EXTRA HEAT</u>
is dispersed
quickly by Blood Stream

Rise in Blood
Temperature
↓ affects
<u>HYPOTHALAMUS</u>

2.
<u>HOT ENVIRONMENTAL
TEMPERATURE</u>
i.e. above BODY
TEMPERATURE (37°C)
*[Body would tend
to gain heat by
RADIATION,
CONVECTION and
CONDUCTION.]*

stimulates

<u>HEAT SENSITIVE</u>
nerve endings
in SKIN

*Reduced
SYMPATHETIC
VASOCONSTRICTOR
tone to*

SECRETOMOTOR

INGOING
NERVE
IMPULSES

OUTGOING
NERVE IMPULSES

BLOOD
VESSELS
DILATED

FAT

1. <u>SKIN BLOOD VESSELS DILATE</u>
*more blood to skin
surface → increased heat
loss from skin by –*
RADIATION
CONVECTION
CONDUCTION
*(Cannot
occur if air
temperature
is above
body's)*
*Increased by
voluntary ingestion of
cold foods and fluids
and by use of fans.*

2. <u>SWEAT GLANDS SECRETE</u>
*Increased heat loss by
evaporation from skin
surface (unless atmosphere
is already water saturated
as e.g. in Tropics)*

3. <u>DIMINISHED HEAT
INSULATION</u>
*by voluntary reduction
of CLOTHING worn*

4. <u>DIMINISHED HEAT
PRODUCTION</u>
SKELETAL MUSCLE *'tone'
reduced – and often
voluntary relaxation
→ less work done →
less heat produced*

5. <u>REDUCTION of 'ENERGY INTAKE'</u>
*by voluntary restriction of
<u>PROTEIN in DIET</u>*

*[The increase in activity during the day
probably accounts for the gradual Physiological
rise in body temperature from about 96.5°F
(35.8°C) in the early morning to about 99.2°F
(37.3°C) in the late afternoon.*

*Unless exercise is very strenuous or environ-
ment is very hot and humid these measures →* <u>RESTORE BODY TEMPERATURE</u>
to NORMAL

MAINTENANCE of BODY TEMPERATURE

Any tendency for the BODY TEMPERATURE to *FALL*

as by

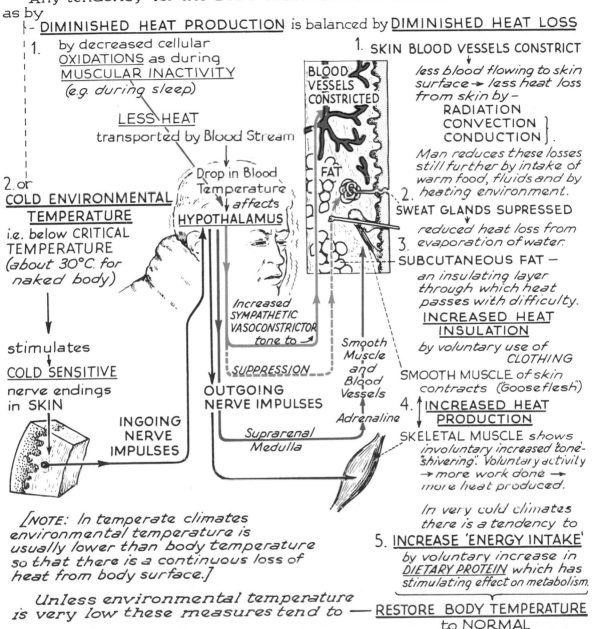

— **DIMINISHED HEAT PRODUCTION** is balanced by **DIMINISHED HEAT LOSS**

1. by decreased cellular
OXIDATIONS as during
MUSCULAR INACTIVITY
(e.g. during sleep)

LESS HEAT
transported by Blood Stream

Drop in Blood
Temperature
affects
HYPOTHALAMUS

2. or
COLD ENVIRONMENTAL
TEMPERATURE
i.e. below CRITICAL
TEMPERATURE
*(about 30°C. for
naked body)*

stimulates
COLD SENSITIVE
nerve endings
in SKIN

INGOING
NERVE
IMPULSES

Increased
SYMPATHETIC
VASOCONSTRICTOR
tone to

SUPPRESSION

OUTGOING
NERVE IMPULSES

*Suprarenal
Medulla*

BLOOD
VESSELS
CONSTRICTED

FAT

*Smooth
Muscle
and
Blood
Vessels*

Adrenaline

1. SKIN BLOOD VESSELS CONSTRICT

*less blood flowing to skin
surface → less heat loss
from skin by –*
RADIATION
CONVECTION
CONDUCTION }.

*Man reduces these losses
still further by intake of
warm food, fluids and by
heating environment.*

2. SWEAT GLANDS SUPRESSED
*reduced heat loss from
evaporation of water.*

3. SUBCUTANEOUS FAT —
*an insulating layer
through which heat
passes with difficulty.*
INCREASED HEAT
INSULATION
*by voluntary use of
CLOTHING*

SMOOTH MUSCLE *of skin
contracts ('Gooseflesh')*

4. ↕ INCREASED HEAT
PRODUCTION
SKELETAL MUSCLE *shows
involuntary increased tone-
'shivering'. Voluntary activity
→ more work done →
more heat produced.*

*In very cold climates
there is a tendency to*

5. INCREASE 'ENERGY INTAKE'
*by voluntary increase in
DIETARY PROTEIN which has
stimulating effect on metabolism.*

[NOTE: *In temperate climates
environmental temperature is
usually lower than body temperature
so that there is a continuous loss of
heat from body surface.*]

*Unless environmental temperature
is very low these measures tend to* — RESTORE BODY TEMPERATURE
to NORMAL

GROWTH

Each individual grows, by repeated cell divisions, from a single cell to a total of 30 trillion or more cells. Growth is most rapid before birth and during 1ˢᵀ year of life.

The proportion of ENERGY INTAKE in food used to build and maintain tissue —

	INFANCY		CHILDHOOD		
	At Birth	3 months	1 year	2 years	9-11 years
proportion of energy intake	40%	40%	20%	20%	4-10%
Average WEIGHT —	3 kg	5 kg	10.4 kg	12.4 kg	27.1 kg
Average HEIGHT —	50 cm	58 cm	73 cm	84 cm	129 cm

A baby is born with epithelia, connective tissues, muscles, nerves and organs all present and formed — but all tissues do not grow at the same rate.

DIFFERENTIAL GROWTH and FUNCTIONAL DEVELOPMENT of TISSUES

↓ *lead to*

CHANGE in BODY PROPORTIONS

— *e.g. Rapid growth of skeletal tissue during childhood. Nervous tissue develops rapidly in first 2 years. Most rapid growth is first at the head then legs begin to lengthen.*

Chiefly PROTEIN being laid down ⟶ or retained

Lymphoid Tissue Growth Spurt 10 YEARS

2 YEARS

1 YEAR

3 MONTHS

At BIRTH

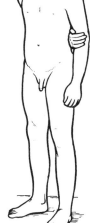

FIRST *(Neutral)* GROWTH PHASE *(Infancy to Puberty)*. No marked difference between sexes – Regulated by Growth-Stimulating Hormone of Anterior Pituitary – stimulates Growth of all Tissues and Organs.

FACTORS INFLUENCING NORMAL GROWTH :
HEREDITY: *To large extent rate of growth/sequence of events is predetermined. Inherited factors control pattern and limitations of growth.*
ENVIRONMENT: *Nutrition: For optimal growth the body requires an adequate and balanced diet.*
ENDOCRINE GLANDS: *All play part: mental as well as physical development is influenced.*

GROWTH

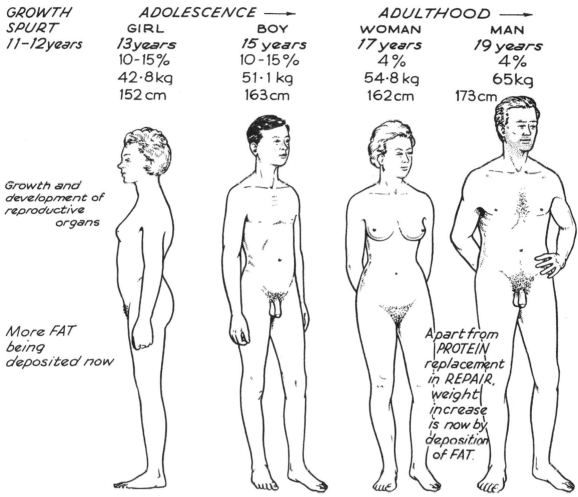

GROWTH SPURT 11-12years	ADOLESCENCE ⟶		ADULTHOOD ⟶	
	GIRL	BOY	WOMAN	MAN
	13 years	15 years	17 years	19 years
	10-15%	10-15%	4%	4%
	42·8 kg	51·1 kg	54·8 kg	65 kg
	152 cm	163 cm	162 cm	173 cm

Growth and
development of
reproductive
         organs

More FAT
being
deposited now

Apart from
PROTEIN
replacement
in REPAIR,
weight
increase
is now by
deposition
of FAT.

SEXUAL GROWTH PHASE *(Puberty to Maturity)*

    Marked difference between sexes is initiated
by Gonad-Stimulating Hormones of Anterior
Pituitary acting on Sex Glands and stimulating
their production of Sex Hormones. These hormones
are largely responsible for development of
secondary sex characteristics and
development of reproductive organs.

REPRODUCTIVE PHASE

---These hormones maintain
secondary sex characteristics
and reproductive ability
during reproductive phase
of adult life.

ENERGY REQUIREMENTS *MALE*

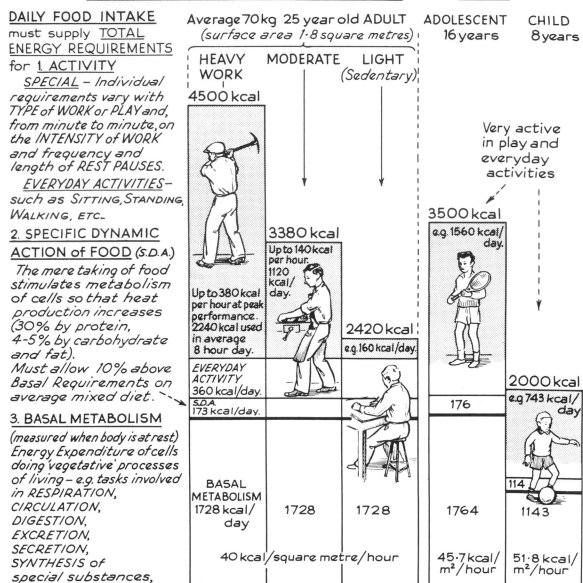

DAILY FOOD INTAKE must supply <u>TOTAL</u> <u>ENERGY REQUIREMENTS</u> for

1. ACTIVITY

SPECIAL – Individual requirements vary with TYPE of WORK or PLAY and, from minute to minute, on the INTENSITY of WORK and frequency and length of REST PAUSES.

EVERYDAY ACTIVITIES – such as SITTING, STANDING, WALKING, ETC.

2. SPECIFIC DYNAMIC ACTION of FOOD (S.D.A.)

The mere taking of food stimulates metabolism of cells so that heat production increases (30% by protein, 4-5% by carbohydrate and fat). Must allow 10% above Basal Requirements on average mixed diet.

3. BASAL METABOLISM

(measured when body is at rest) Energy Expenditure of cells doing 'vegetative' processes of living – e.g. tasks involved in RESPIRATION, CIRCULATION, DIGESTION, EXCRETION, SECRETION, SYNTHESIS of special substances, KEEPING BODY TEMPERATURE at 37°C., GROWTH and REPAIR.

Average 70 kg 25 year old ADULT
(surface area 1·8 square metres)

ADOLESCENT 16 years   CHILD 8 years

HEAVY WORK   MODERATE   LIGHT (Sedentary)

Very active in play and everyday activities

4500 kcal

Up to 380 kcal per hour at peak performance. 2240 kcal used in average 8 hour day.

EVERYDAY ACTIVITY 360 kcal/day.

S.D.A. 173 kcal/day.

BASAL METABOLISM 1728 kcal/day

3380 kcal

Up to 140 kcal per hour. 1120 kcal/day.

2420 kcal

e.g. 160 kcal/day.

3500 kcal

e.g. 1560 kcal/day.

176

2000 kcal

e.g. 743 kcal/day

114

HEAVY WORK	MODERATE	LIGHT	ADOLESCENT	CHILD
1728	1728	1728	1764	1143

40 kcal/square metre/hour   45·7 kcal/m²/hour   51·8 kcal/m²/hour

Basal metabolism is proportional to surface area of body and varies with sex and age.

[All figures are approximate and intended only as a general guide.]

ENERGY REQUIREMENTS *FEMALE*

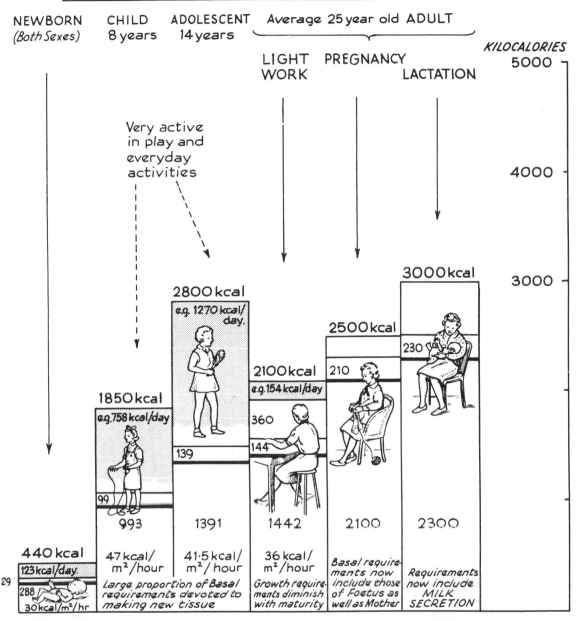

NEWBORN (Both Sexes)   CHILD 8 years   ADOLESCENT 14 years   Average 25 year old ADULT   KILOCALORIES

5000

LIGHT WORK   PREGNANCY   LACTATION

4000

Very active in play and everyday activities

3000

3000 kcal

2800 kcal
e.g. 1270 kcal/day

2500 kcal

230

2100 kcal
e.g.154 kcal/day

210

1850 kcal
e.g.758 kcal/day

360

144

139

99

993   1391   1442   2100   2300

440 kcal
123 kcal/day
29
288
30 kcal/m²/hr

47 kcal/ m²/hour
Large proportion of Basal requirements devoted to making new tissue

41·5 kcal/ m²/hour

36 kcal/ m²/hour
Growth require-ments diminish with maturity

Basal require-ments now include those of Foetus as well as Mother

Requirements now include MILK SECRETION

[Proportion needed for growth diminishes in both sexes with age.]

49

BALANCED DIET

The individual's daily energy requirements are best obtained by eating well-balanced meals which contain the essential *body-building* and *protective* foods as well as *energy-giving* foods.

1st. Class Proteins

Milk & Milk Foods
Vit.B₂
Ca
P
Vit.D   Nacl

FROM EACH GROUP:—

Children:
3–4 servings per day.
Adults:
2 or more servings per day.

1 – 2 servings per day.

Adult total protein requirement averages 70 to 100 grams/day.

1st. Class Proteins

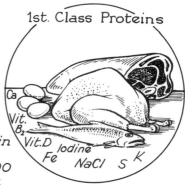

Ca
Vit. B₂
Vit.D   Iodine
Fe   NaCl   S   K

2nd. Class Proteins
(with some carbohydrates)

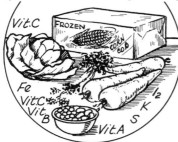

Vit.C
Fe
Vit.C
Vit.B
Vit.A
K
I₂
S

1 serving per day

1 serving per day

NOTE:— *The normal daily requirement of vitamins and essential minerals is met by such a balanced diet. In addition to fluid in these foods, daily* Water *requirement is approximately 1 litre. This varies with sweat loss, etc. (see Index under "Water Balance")*

To Provide chiefly Vitamin C

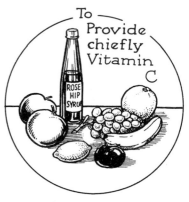

Chiefly for their carbohydrates

Vit.B
Fe
(Vit.C)

4 or more servings per day

For Fats *and* Fat Soluble *Vitamins*

(NaCl)

Some per day

CHAPTER 3

DIGESTIVE SYSTEM

DIGESTIVE SYSTEM

The ALIMENTARY CANAL and ASSOCIATED GLANDS } Special System for dealing with FOOD and FLUIDS

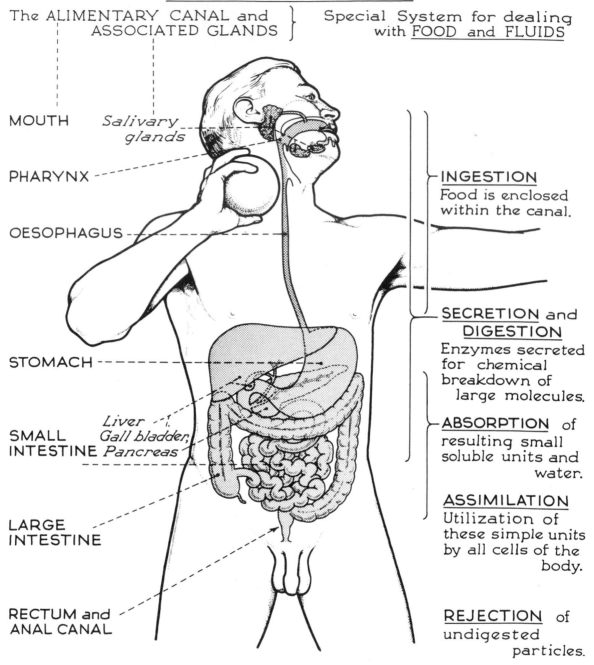

MOUTH

Salivary glands

PHARYNX

OESOPHAGUS

STOMACH

*Liver
Gall bladder,*
SMALL
INTESTINE *Pancreas*

LARGE
INTESTINE

RECTUM and
ANAL CANAL

INGESTION
Food is enclosed
within the canal.

SECRETION and
DIGESTION
Enzymes secreted
for chemical
breakdown of
large molecules.

ABSORPTION of
resulting small
soluble units and
water.

ASSIMILATION
Utilization of
these simple units
by all cells of the
body.

REJECTION of
undigested
particles.

52

PROGRESS of FOOD along ALIMENTARY CANAL

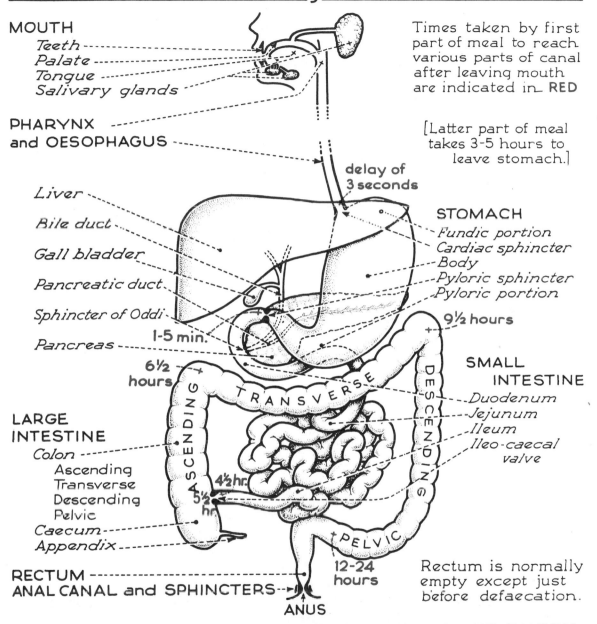

MOUTH
  Teeth
  Palate
  Tongue
  Salivary glands

PHARYNX and OESOPHAGUS

Times taken by first part of meal to reach various parts of canal after leaving mouth are indicated in **RED**

[Latter part of meal takes 3-5 hours to leave stomach.]

delay of 3 seconds

Liver

Bile duct

Gall bladder

Pancreatic duct

Sphincter of Oddi

1-5 min.

Pancreas

6½ hours

STOMACH
  Fundic portion
  Cardiac sphincter
  Body
  Pyloric sphincter
  Pyloric portion

9½ hours

SMALL INTESTINE
  Duodenum
  Jejunum
  Ileum
  Ileo-caecal valve

TRANSVERSE

ASCENDING

DESCENDING

LARGE INTESTINE
  Colon
    Ascending
    Transverse
    Descending
    Pelvic
  Caecum
  Appendix

4½ hr.

5½ hr

PELVIC

12-24 hours

RECTUM
ANAL CANAL and SPHINCTERS

ANUS

Rectum is normally empty except just before defaecation.

During its progress along the canal FOOD is subjected to MECHANICAL as well as CHEMICAL changes to render it suitable for absorption and assimilation.

DIGESTION in the MOUTH

MECHANICAL PROCESSES

MASTICATION

Chewing movements of TEETH, TONGUE, CHEEKS, LIPS, LOWER JAW, break down food, mix it with SALIVA and roll it into a moist soft mass (BOLUS) suitable for SWALLOWING

[Mastication is under VOLUNTARY control]

PAROTID SALIVARY GLAND (serous)

SEROUS ACINI

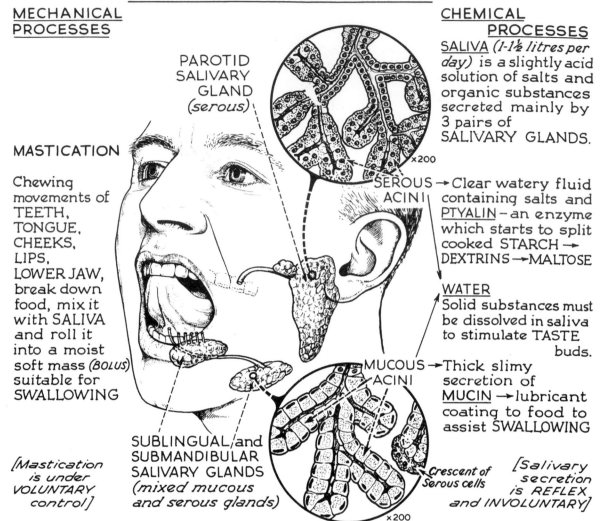

×200

SUBLINGUAL and SUBMANDIBULAR SALIVARY GLANDS (mixed mucous and serous glands)

MUCOUS ACINI

Crescent of Serous cells

×200

CHEMICAL PROCESSES

SALIVA (*1-1½ litres per day*) is a slightly acid solution of salts and organic substances secreted mainly by 3 pairs of SALIVARY GLANDS.

→ Clear watery fluid containing salts and PTYALIN – an enzyme which starts to split cooked STARCH → DEXTRINS → MALTOSE

WATER
Solid substances must be dissolved in saliva to stimulate TASTE buds.

→ Thick slimy secretion of MUCIN → lubricant coating to food to assist SWALLOWING

[Salivary secretion is REFLEX and INVOLUNTARY]

Other important (*non-digestive*) functions of SALIVA:-

CLEANSING — Mouth and teeth kept free of debris, etc., to inhibit bacteria.

MOISTENING and LUBRICATING — Soft parts of mouth kept pliable for SPEECH.

EXCRETORY — Many organic substances (*e.g. urea, sugar*) and inorganic substances (*e.g. mercury, lead*) can be excreted in saliva.

CONTROL of SALIVARY SECRETION

Increased secretion at mealtimes is REFLEX *(involuntary)*.

Salivary Reflexes are of two types:–

(a) <u>UNCONDITIONED</u> *(inborn)*

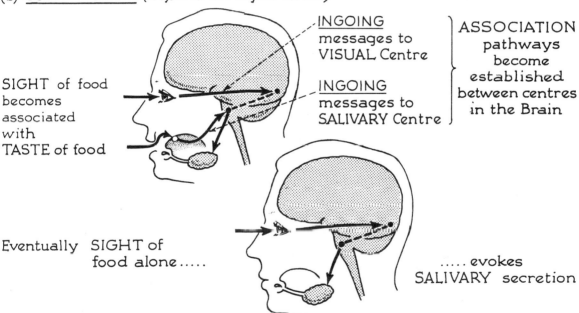

INGOING messages along TASTE nerves to Controlling SALIVARY Centre in MEDULLA OBLONGATA

OUTGOING messages in SECRETOMOTOR nerves from the AUTONOMIC NERVOUS SYSTEM Centre to glands evoke secretion of SALIVA.

FOOD stimulates nerve endings in mouth

(b) <u>CONDITIONED</u> *(depend on experience)*

SIGHT of food becomes associated with TASTE of food

INGOING messages to VISUAL Centre

INGOING messages to SALIVARY Centre

ASSOCIATION pathways become established between centres in the Brain

Eventually SIGHT of food alone.....

..... evokes SALIVARY secretion

Similar conditioned reflexes are established by smell, by thought of food, and even by the sounds of its preparation.

OESOPHAGUS

The Oesophagus is a muscular tube about 25 cm. long which conveys ingested food and fluid from the Mouth to the Stomach.

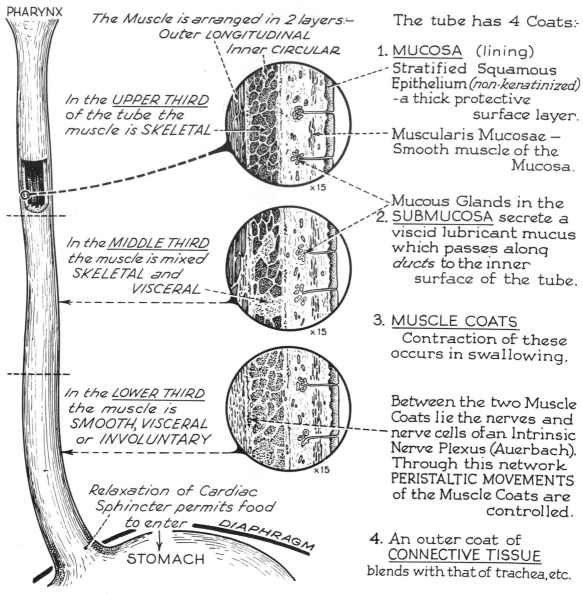

PHARYNX

The Muscle is arranged in 2 layers:—
Outer LONGITUDINAL
Inner CIRCULAR

In the UPPER THIRD of the tube the muscle is SKELETAL

In the MIDDLE THIRD the muscle is mixed SKELETAL and VISCERAL

In the LOWER THIRD the muscle is SMOOTH, VISCERAL or INVOLUNTARY

Relaxation of Cardiac Sphincter permits food to enter

DIAPHRAGM

STOMACH

×15

×15

×15

The tube has 4 Coats:—

1. MUCOSA (lining)
Stratified Squamous Epithelium (*non-keratinized*) – a thick protective surface layer.

Muscularis Mucosae — Smooth muscle of the Mucosa.

Mucous Glands in the
2. SUBMUCOSA secrete a viscid lubricant mucus which passes along *ducts* to the inner surface of the tube.

3. MUSCLE COATS
Contraction of these occurs in swallowing.

Between the two Muscle Coats lie the nerves and nerve cells of an Intrinsic Nerve Plexus (Auerbach). Through this network PERISTALTIC MOVEMENTS of the Muscle Coats are controlled.

4. An outer coat of CONNECTIVE TISSUE blends with that of trachea, etc.

Except during passage of food the oesophagus is flattened and closed; its mucosa thrown into several longitudinal folds.

56

SWALLOWING

Swallowing is a complex act initiated *voluntarily* and completed *involuntarily (or reflexly)*

STIMULUS	INGOING PATHWAY	CENTRE IN MEDULLA OBLONGATA	OUTGOING PATHWAY	EFFECT

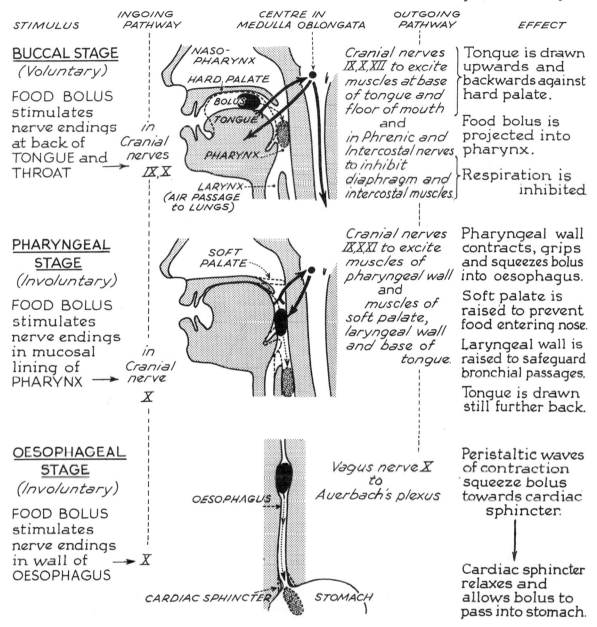

BUCCAL STAGE
(Voluntary)

FOOD BOLUS stimulates nerve endings at back of TONGUE and THROAT

in Cranial nerves IX, X →

NASO-PHARYNX
HARD PALATE
BOLUS
TONGUE
PHARYNX
LARYNX (AIR PASSAGE TO LUNGS)

Cranial nerves IX, X, XII to excite muscles at base of tongue and floor of mouth and in Phrenic and Intercostal nerves to inhibit diaphragm and intercostal muscles.

Tongue is drawn upwards and backwards against hard palate.

Food bolus is projected into pharynx.

Respiration is inhibited.

PHARYNGEAL STAGE
(Involuntary)

FOOD BOLUS stimulates nerve endings in mucosal lining of PHARYNX →

in Cranial nerve X

SOFT PALATE

Cranial nerves IX, X, XI to excite muscles of pharyngeal wall and muscles of soft palate, laryngeal wall and base of tongue.

Pharyngeal wall contracts, grips and squeezes bolus into oesophagus.

Soft palate is raised to prevent food entering nose.

Laryngeal wall is raised to safeguard bronchial passages.

Tongue is drawn still further back.

OESOPHAGEAL STAGE
(Involuntary)

FOOD BOLUS stimulates nerve endings in wall of OESOPHAGUS → X

OESOPHAGUS
CARDIAC SPHINCTER
STOMACH

Vagus nerve X to Auerbach's plexus

Peristaltic waves of contraction squeeze bolus towards cardiac sphincter.

Cardiac sphincter relaxes and allows bolus to pass into stomach.

57

STOMACH

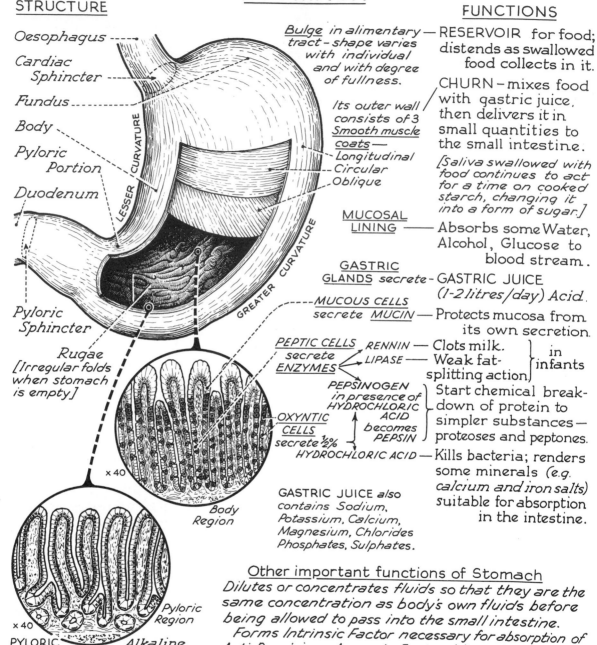

STRUCTURE

- Oesophagus
- Cardiac Sphincter
- Fundus
- Body
- Pyloric Portion
- Duodenum
- Pyloric Sphincter
- Rugae *[Irregular folds when stomach is empty]*

LESSER CURVATURE

GREATER CURVATURE

Bulge in alimentary tract – shape varies with individual and with degree of fullness.

Its outer wall consists of 3 Smooth muscle coats —
- Longitudinal
- Circular
- Oblique

× 40

Body Region

× 40

PYLORIC GLANDS

Pyloric Region

Alkaline Mucus

FUNCTIONS

RESERVOIR for food; distends as swallowed food collects in it.

CHURN – mixes food with gastric juice, then delivers it in small quantities to the small intestine.

[Saliva swallowed with food continues to act for a time on cooked starch, changing it into a form of sugar.]

MUCOSAL LINING — Absorbs some Water, Alcohol, Glucose to blood stream.

GASTRIC GLANDS secrete – GASTRIC JUICE (1–2 litres/day) Acid.

MUCOUS CELLS secrete MUCIN — Protects mucosa from its own secretion.

PEPTIC CELLS secrete ENZYMES
- RENNIN — Clots milk. } in infants
- LIPASE — Weak fat-splitting action. }

PEPSINOGEN in presence of HYDROCHLORIC ACID becomes PEPSIN } Start chemical breakdown of protein to simpler substances — proteoses and peptones.

OXYNTIC CELLS secrete ½% HYDROCHLORIC ACID — Kills bacteria; renders some minerals (e.g. calcium and iron salts) suitable for absorption in the intestine.

GASTRIC JUICE also contains Sodium, Potassium, Calcium, Magnesium, Chlorides Phosphates, Sulphates.

Other important functions of Stomach

Dilutes or concentrates fluids so that they are the same concentration as body's own fluids before being allowed to pass into the small intestine.

Forms Intrinsic Factor necessary for absorption of Anti-Pernicious Anaemia Factor, Vitamin B_{12}.

GASTRIC JUICE

SECRETION of Gastric Juice is under 2 types of CONTROL:-

(a) NERVOUS - Messages are conveyed rapidly from Brain Centre by Nerve Fibres of the AUTONOMIC NERVOUS SYSTEM for *immediate* effect. e.g. stimulation of Parasympathetic (Vagus) nerves to GASTRIC GLANDS → secretion of HIGHLY ACID JUICE containing ENZYMES.

(b) HUMORAL - Message is CHEMICAL and carried in BLOOD STREAM for *slower* and *longer-lasting* control. [Note:- *Chemical messengers (Hormones) travel all over the body even though the message stimulates only one part — e.g. in this case the Gastric Glands.*]

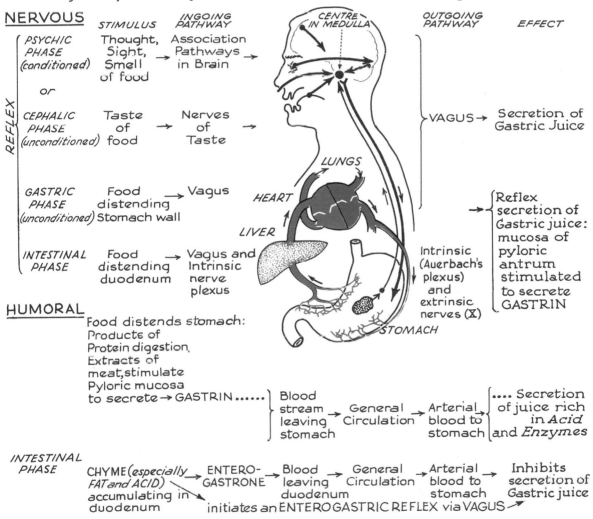

NERVOUS

| | STIMULUS | INGOING PATHWAY | CENTRE IN MEDULLA | OUTGOING PATHWAY | EFFECT |

REFLEX

PSYCHIC PHASE (conditioned) — Thought, Sight, Smell of food → Association Pathways in Brain →

or

CEPHALIC PHASE (unconditioned) — Taste of food → Nerves of Taste →

VAGUS → Secretion of Gastric Juice

GASTRIC PHASE (unconditioned) — Food distending Stomach wall → Vagus

INTESTINAL PHASE — Food distending duodenum → Vagus and Intrinsic nerve plexus

Intrinsic (Auerbach's plexus) and extrinsic nerves (X) → Reflex secretion of Gastric juice: mucosa of pyloric antrum stimulated to secrete GASTRIN

LUNGS
HEART
LIVER
STOMACH

HUMORAL

Food distends stomach: Products of Protein digestion, Extracts of meat, stimulate Pyloric mucosa to secrete → GASTRIN ...... } Blood stream leaving stomach → General Circulation → Arterial blood to stomach { .... Secretion of juice rich in *Acid* and *Enzymes*

INTESTINAL PHASE — CHYME (*especially FAT and ACID*) accumulating in duodenum → ENTERO-GASTRONE → Blood leaving duodenum → General Circulation → Arterial blood to stomach → Inhibits secretion of Gastric juice

initiates an ENTEROGASTRIC REFLEX via VAGUS ↗

MOVEMENTS of the STOMACH

Very little movement is seen in empty stomach until onset of HUNGER.

FILLING

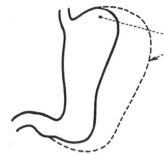

FUNDUS and
GREATER CURVATURE bulge and lengthen as stomach fills with food. This is "Receptive Relaxation" – the smooth muscle cells relax so that the pressure in the stomach does not rise until the organ is fairly large.

TWO TYPES of MOVEMENT occur while food is in stomach.

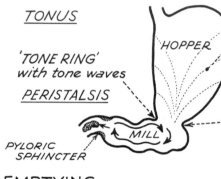

TONUS

'TONE RING' with tone waves

PERISTALSIS

HOPPER

MILL

PYLORIC SPHINCTER

BODY – Food lies in layers and the walls exert a slight but steady pressure on it. This squeezes food steadily towards the PYLORUS even when stomach is becoming relatively empty near the end of a meal.

From INCISURA ANGULARIS vigorous waves of contraction mix food with digestive juices and carry chyme through the normally relaxed PYLORUS into the first part of the duodenum.

EMPTYING

CONTROL of MOTILITY and EMPTYING

1. Enterogastric
   NERVOUS Reflex
      As Chyme enters and distends duodenum
   HUMORAL

   inhibition of VAGUS→

   release of ENTEROGASTRONE→ to Blood Stream

   Inhibits tone and peristalsis temporarily.
   ↓
   Gastric motility depressed.
   ↓
   Temporarily slows emptying of stomach.

2.
      As duodenum empties

   stimulation of VAGUS →

   withdrawal of ENTEROGASTRONE→

   Waves of peristalsis become stronger.
   ↓
   Gastric motility increased.
   ↓
   Emptying speeded up again.

   This mechanism is important in regulating the rate of emptying of the stomach from moment to moment during the digestion of a meal.

[STARCHY FOODS leave the stomach quickly: MEAT leaves relatively slowly: FATTY FOODS pass through most slowly of all.]

60

VOMITING

Vomiting is a <u>REFLEX</u> act.

Nerve pathways involved:-

STIMULUS → INGOING NERVE PATHWAY → VOMITING CENTRE IN MEDULLA OBLONGATA → OUTGOING NERVE PATHWAY → EFFECT

<u>PSYCHIC INFLUENCES</u>
e.g. fear, anxiety, disgust.

<u>DISCORDANT INGOING IMPULSES</u>
<u>EYES</u>
<u>LABYRINTH of EARS</u>
e.g. stimulation of nerve endings of utricle in Motion Sickness.

<u>LOCAL IRRITATION</u> of
nerve endings
(parasympathetic and sympathetic)
by drugs, poisons, or in
disease in PHARYNX,
OESOPHAGUS, STOMACH,
INTESTINE *(especially*
APPENDIX), GALL
BLADDER, UTERUS, in
FAILING HEART or from
pain endings in any
organ of the body.

<u>BLOOD BORNE EFFECTS</u>
Certain EMETIC DRUGS,
e.g. Apomorphine, lower
threshold of Centre
to stimulation.
<u>METABOLIC</u>
<u>DISTURBANCES</u>
e.g. in Pregnancy and
Fatigue

VAGUS — *Oesophagus*
— *Cardiac sphincter*
— *Body of stomach* ⎱ → Relax.
— *Pyloric part* → Contracts strongly.

SPINAL NERVES — *Abdominal muscles* → Contract.

PHRENIC NERVE — *Diaphragm* → Descends to compress abominal cavity.

[Larynx is raised to protect air passages.]

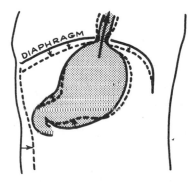

As a result of these changes the <u>Stomach is compressed</u> with <u>PASSIVE emptying of contents</u>.

PANCREAS

The Pancreas is a large gland lying across the Posterior Abdominal Wall. It has 2 secretions – a *digestive* secretion poured into the duodenum, and a *hormonal* secretion passed into the blood stream.

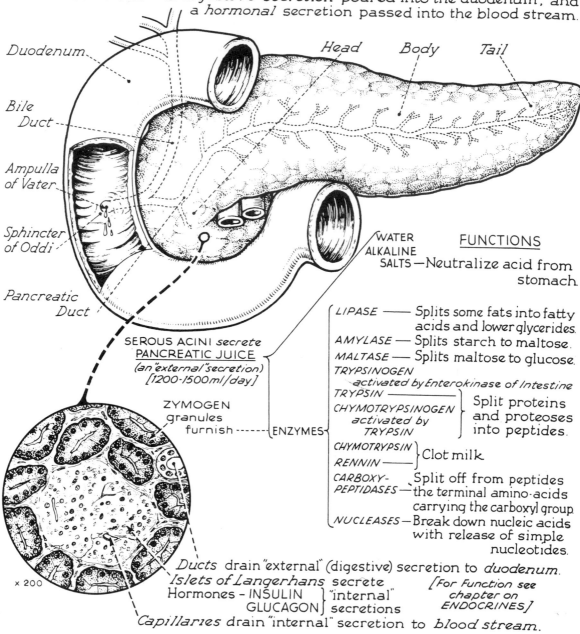

Duodenum

Bile Duct

Ampulla of Vater

Sphincter of Oddi

Pancreatic Duct

Head   Body   Tail

SEROUS ACINI *secrete*
PANCREATIC JUICE
(an "external" secretion)
[1200-1500 ml/day]

ZYMOGEN
granules
furnish

× 200

WATER
ALKALINE
SALTS —Neutralize acid from stomach.

FUNCTIONS

ENZYMES

LIPASE —— Splits some fats into fatty acids and lower glycerides.

AMYLASE — Splits starch to maltose.

MALTASE — Splits maltose to glucose.

TRYPSINOGEN activated by Enterokinase of Intestine
TRYPSIN ——
CHYMOTRYPSINOGEN activated by TRYPSIN
} Split proteins and proteoses into peptides.

CHYMOTRYPSIN
RENNIN ——
} Clot milk

CARBOXY-PEPTIDASES — Split off from peptides the terminal amino-acids carrying the carboxyl group.

NUCLEASES — Break down nucleic acids with release of simple nucleotides.

Ducts drain "external" (digestive) secretion to *duodenum*.

Islets of Langerhans secrete
Hormones - INSULIN
GLUCAGON } "internal" secretions

[For Function see chapter on ENDOCRINES]

Capillaries drain "internal" secretion to *blood stream*.

PANCREATIC JUICE

SECRETION of Pancreatic Juice is under 2 types of CONTROL :- NERVOUS and HUMORAL. The humoral mechanism is the more important.

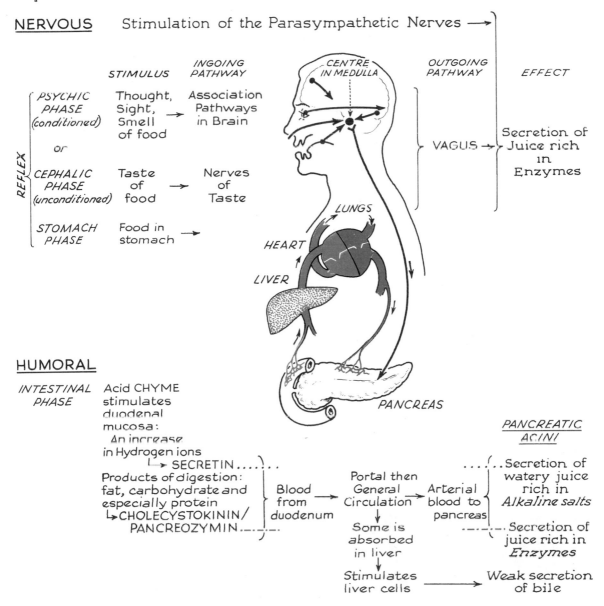

NERVOUS   Stimulation of the Parasympathetic Nerves ⟶

	STIMULUS	INGOING PATHWAY		OUTGOING PATHWAY	EFFECT

REFLEX

PSYCHIC PHASE (conditioned)   Thought, Sight, Smell of food   →   Association Pathways in Brain

or

CEPHALIC PHASE (unconditioned)   Taste of food   →   Nerves of Taste

STOMACH PHASE   Food in stomach   →

CENTRE IN MEDULLA

VAGUS → Secretion of Juice rich in Enzymes

LUNGS

HEART

LIVER

HUMORAL

INTESTINAL PHASE   Acid CHYME stimulates duodenal mucosa:
An increase in Hydrogen ions
⤷ SECRETIN......

Products of digestion: fat, carbohydrate and especially protein
⤷ CHOLECYSTOKININ/ PANCREOZYMIN.

Blood from duodenum

Portal then General Circulation →

Some is absorbed in liver
↓
Stimulates liver cells   ⟶   Weak secretion of bile

Arterial blood to pancreas

PANCREAS

PANCREATIC ACINI

......Secretion of watery juice rich in *Alkaline salts*

Secretion of juice rich in *Enzymes*

LIVER and GALL BLADDER

The Liver is a large highly complex organ with many functions. One of these is the production of BILE (500-1000 ml/day)

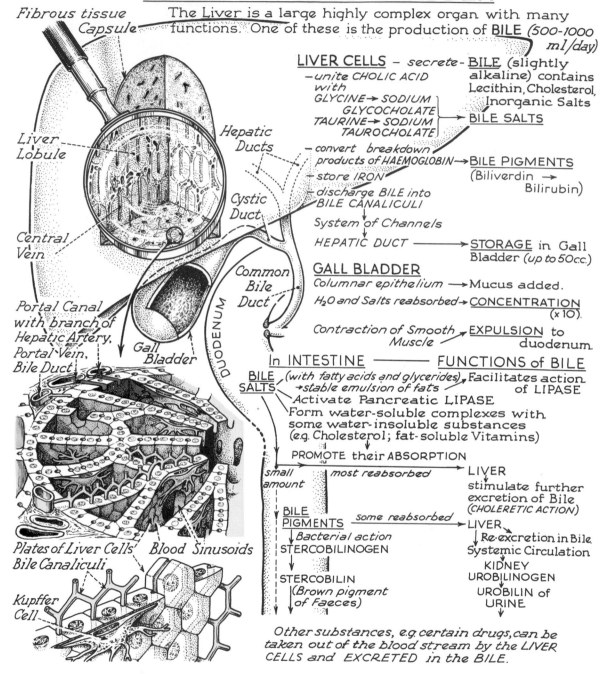

Fibrous tissue Capsule

Liver Lobule

Central Vein

Portal Canal with branch of Hepatic Artery, Portal Vein, Bile Duct

Hepatic Ducts

Cystic Duct

Common Bile Duct

Gall Bladder

DUODENUM

Plates of Liver Cells *Blood Sinusoids*
Bile Canaliculi

Kupffer Cell

LIVER CELLS – *secrete* - BILE (slightly alkaline) contains Lecithin, Cholesterol, Inorganic Salts
– *unite CHOLIC ACID with*
GLYCINE → SODIUM GLYCOCHOLATE
TAURINE → SODIUM TAUROCHOLATE } → BILE SALTS

– *convert breakdown products of HAEMOGLOBIN* → BILE PIGMENTS (Biliverdin → Bilirubin)
– *store IRON*
– *discharge BILE into BILE CANALICULI*

System of Channels

HEPATIC DUCT ⟶ STORAGE in Gall Bladder (up to 50cc.)

GALL BLADDER
Columnar epithelium ⟶ Mucus added.
H_2O and Salts reabsorbed → CONCENTRATION (x 10).

Contraction of Smooth Muscle ⟶ EXPULSION to duodenum.

In INTESTINE ⟶ FUNCTIONS of BILE

BILE SALTS (with fatty acids and glycerides) → Facilitates action of LIPASE
→ stable emulsion of fats
Activate Pancreatic LIPASE
Form water-soluble complexes with some water-insoluble substances (e.g. Cholesterol; fat-soluble Vitamins)

PROMOTE their ABSORPTION

small amount   most reabsorbed ⟶ LIVER stimulate further excretion of Bile (CHOLERETIC ACTION)

BILE PIGMENTS   some reabsorbed ⟶ LIVER
↓ Bacterial action   ↓ Re-excretion in Bile
STERCOBILINOGEN   Systemic Circulation
↓   KIDNEY
STERCOBILIN   UROBILINOGEN
(Brown pigment of Faeces)   UROBILIN of URINE
        ↓

Other substances, e.g. certain drugs, can be taken out of the blood stream by the LIVER CELLS and EXCRETED in the BILE.

EXPULSION of BILE *(from the Gall Bladder)*

BILE is SECRETED continuously by the LIVER.
It is STORED and CONCENTRATED in the GALL BLADDER.
Periodically *(e.g. during a meal)* it is DISCHARGED into the DUODENUM.
NERVOUS and HUMORAL factors influence this expulsion:-

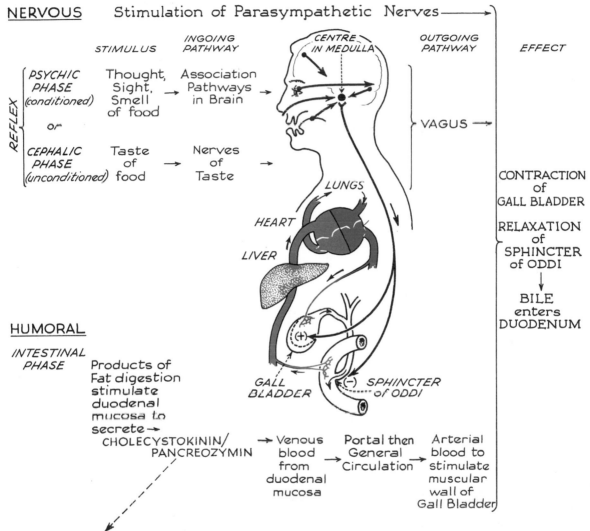

NERVOUS   Stimulation of Parasympathetic Nerves ⟶

| | *STIMULUS* | *INGOING PATHWAY* | *CENTRE IN MEDULLA* | *OUTGOING PATHWAY* | *EFFECT* |

REFLEX

PSYCHIC PHASE (conditioned) — Thought, Sight, Smell of food → Association Pathways in Brain →

or

CEPHALIC PHASE (unconditioned) — Taste of food → Nerves of Taste →

⟶ VAGUS ⟶

CONTRACTION of GALL BLADDER

RELAXATION of SPHINCTER of ODDI

↓

BILE enters DUODENUM

LUNGS

HEART

LIVER

GALL BLADDER

(+)

(−) SPHINCTER of ODDI

HUMORAL

INTESTINAL PHASE — Products of Fat digestion stimulate duodenal mucosa to secrete → CHOLECYSTOKININ/ PANCREOZYMIN → Venous blood from duodenal mucosa → Portal then General Circulation → Arterial blood to stimulate muscular wall of Gall Bladder

[This hormone is probably the main factor controlling Gall Bladder
contraction in man]

SMALL INTESTINE

The Small Intestine is a long muscular tube - over 6 metres in length. It receives CHYME in small quantities from the STOMACH; PANCREATIC JUICE from the PANCREAS; BILE from the GALL BLADDER.

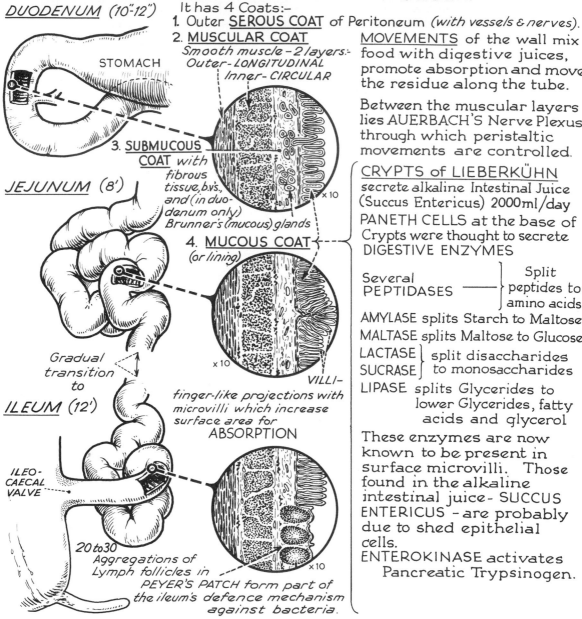

DUODENUM (10"-12")

STOMACH

It has 4 Coats:-
1. Outer **SEROUS COAT** of Peritoneum *(with vessels & nerves).*
2. **MUSCULAR COAT**
   Smooth muscle - 2 layers:-
   Outer - LONGITUDINAL
   Inner - CIRCULAR

3. **SUBMUCOUS COAT** *with fibrous tissue, b.v's, and (in duodenum only) Brunner's (mucous) glands*

JEJUNUM (8')

4. **MUCOUS COAT** *(or lining)*

Gradual transition to

ILEUM (12')

ILEO-CAECAL VALVE

× 10

VILLI-finger-like projections with microvilli which increase surface area for ABSORPTION

× 10

20 to 30 Aggregations of Lymph follicles in PEYER'S PATCH *form part of the ileum's defence mechanism against bacteria.*

× 10

MOVEMENTS of the wall mix food with digestive juices, promote absorption and move the residue along the tube.

Between the muscular layers lies AUERBACH'S Nerve Plexus through which peristaltic movements are controlled.

CRYPTS of LIEBERKÜHN secrete alkaline Intestinal Juice (Succus Entericus) 2000ml/day
PANETH CELLS at the base of Crypts were thought to secrete DIGESTIVE ENZYMES

Several PEPTIDASES — } Split peptides to amino acids

AMYLASE splits Starch to Maltose
MALTASE splits Maltose to Glucose
LACTASE } split disaccharides
SUCRASE } to monosaccharides
LIPASE splits Glycerides to lower Glycerides, fatty acids and glycerol

These enzymes are now known to be present in surface microvilli. Those found in the alkaline intestinal juice - SUCCUS ENTERICUS - are probably due to shed epithelial cells.
ENTEROKINASE activates Pancreatic Trypsinogen.

The BASIC PATTERN of the GUT WALL

The wall of the digestive tube shows a basic structural pattern.
FOUR coats are seen in transverse section.

STRUCTURE

1. MUCOUS COAT *(or MUCOSA)*

SURFACE EPITHELIUM *[type varies with*
with its GLANDS *site and function]*

LOOSE FIBROUS TISSUE *(lamina propria)*
with capillaries and lymphatic vessels.

MUSCULARIS MUCOSAE
(thin layers of smooth
muscle)

LYMPHATIC TISSUE

2. SUBMUCOUS COAT
(or SUBMUCOSA)

DENSE FIBROUS TISSUE
in which lie bloodvessels,
lymphatic vessels and
MEISSNER'S NERVE PLEXUS
[Glands in oesophagus
and 1st part of duodenum]
LYMPHATIC TISSUE

3. MUSCULAR COAT
(or MUSCULARIS EXTERNA)

SMOOTH MUSCLE LAYERS
Inner-*Circular*
arrangement
Outer - *Longitudinal*
arrangement
Between them, blood- and
lymphatic vessels and
AUERBACH'S NERVE PLEXUS
[Some striated muscle
in oesophagus]

(After GARVEN)

4. SEROUS COAT *(or SEROSA)*

FIBROUS TISSUE
with fat, blood- and lymphatic vessels
MESOTHELIUM
Where tube is suspended by a MESENTERY
the serosa is formed by visceral layer of Peritoneum

[Where there is no mesentery — replaced by
fibrous tissue which merges with
surrounding fibrous tissue]

FUNCTIONS

Layer in close contact with gut
contents: specialized for —
SECRETION
ABSORPTION - nutrients and
hormones to blood stream.

MOBILITY (can continually change
degree of folding and the surface
of contact with gut contents).

DEFENCE against bacteria.

STRONG LAYER of SUPPLY
to specialized mucosa
(Bloodvessels supply
needs and remove
absorbed materials).

CO-ORDINATION of motor
and secretory activities
of mucosa.

MOVEMENT

Controls diameter of tube.
Mixes contents.
Propagates contents
along tube.

Nervous elements
co-ordinate secretory and
muscular activities.

Carries nerves, blood- and
lymphatic vessels to and
from mesentery.

Forms smooth, moist
membrane which reduces
friction between contacting
surfaces in the peritoneal
cavity.

INTESTINAL SECRETIONS

Two types of secretion are found in the intestine *(about 2 litres per day)*.

1. <u>ALKALINE</u> secretion (with mucus) secreted by Submucosal BRUNNER'S GLANDS in the first part of the Duodenum to protect the mucosa against ACID of the Stomach.

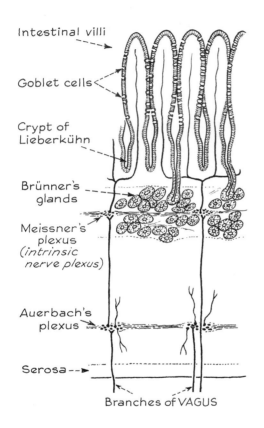

Intestinal villi

Goblet cells

Crypt of Lieberkühn

Brünner's glands

Meissner's plexus *(intrinsic nerve plexus)*

Auerbach's plexus

Serosa

Branches of VAGUS

<u>*STIMULUS for secretion*</u>

(a) Tactile or Irritating stimuli of the overlying mucosa via Intrinsic nerve plexus.

(b) Vagal stimulation

↓

secretion concurrent with increase in stomach secretion.

(c) Intestinal hormones, especially secretin.

[(d) Brünner's glands are inhibited by sympathetic stimulation.]

NOTE: Mucus is also secreted by goblet cells in the intestinal mucosa.

2. <u>SUCCUS ENTERICUS</u> secreted by Crypts of Lieberkühn.
   Its enzymes are probably due to the presence of shed epithelial cells with the enzymes contained in their surface microvilli.

MOVEMENTS of the SMALL INTESTINE

The Duodenum receives food in small quantities from the Stomach. The mixture of food and digestive juices — Chyme - is passed along the length of the Small Intestine.

TWO TYPES of MOVEMENT

SEGMENTATION —
Rhythmical alternating contractions and relaxations

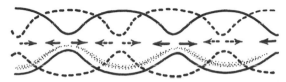

— These "shuttling" movements serve to mix CHYME and to bring it into contact with the absorptive mucosa, i.e. DIGESTION and ABSORPTION are promoted.

This type of movement is MYOGENIC, i.e. it is the property of the smooth muscle cells. It does not depend on a nervous mechanism.

PERISTALSIS —
Food probably acts as stimulus to stretch receptors in muscle and perhaps in mucous membrane

Circular muscle behind bolus CONTRACTS   Muscle in front of bolus RELAXES

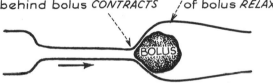

— Waves of this contraction move the food along the canal.

The contraction behind the bolus sweeps it into the relaxed portion of the tube ahead.

This type of movement is NEUROGENIC, i.e. it is carried out through a "Local" Reflex mediated through INTRINSIC nerve plexuses within the wall of the tube.

The Extrinsic nerve supply influences it. { Parasympathetic stimulation ⟶ ↑contractions.
Sympathetic stimulation ⟶ ↓motility.

EMPTYING

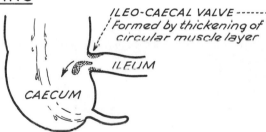

ILEO-CAECAL VALVE -------- Formed by thickening of circular muscle layer

ILEUM

CAECUM

opens and closes during digestion to allow spurts of fluid material from the Ileum to enter the Large Intestine.

This is a "Central" Reflex initiated when food enters the stomach and carried out through EXTRINSIC nerves to wall of tube.

Meals of different composition travel along the intestine at different rates. DIGESTION and ABSORPTION of food are usually complete by the time the residue reaches the Ileo-caecal valve.

ABSORPTION in SMALL INTESTINE

Absorption of most digested foodstuffs occurs in the Small Intestine through the Striated Border Epithelium covering the Villi.

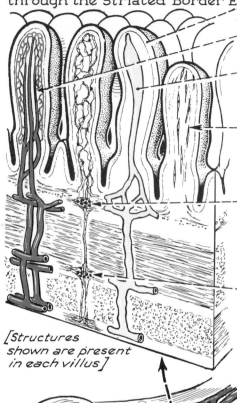

Amino Acids; Sugars; Minerals; Glycerol; some Fatty Acids and Vitamins } into CAPILLARIES

Glycerides and some Fatty Acids and Fat-soluble Vitamins } into LACTEALS

Absorption is aided by movements of villi ("Villus Pump" mechanism) brought about by contraction of smooth muscle – extensions of muscularis mucosae – present in core of villus and attached to lacteal and to basement membrane of epithelium.

Movements of the villi are regulated through NERVE PLEXUS of MEISSNER and perhaps stimulated by intestinal hormones.

Movements of the muscular coats are regulated through the NERVE PLEXUS of AUERBACH.

The Intestines are suspended from the Posterior Abdominal wall by a delicate membrane – the MESENTERY – which carries the mesenteric arteries and veins to and from the gut; and lymphatic vessels on their way via the mesenteric lymph nodes to Thoracic Duct.

[Structures shown are present in each villus]

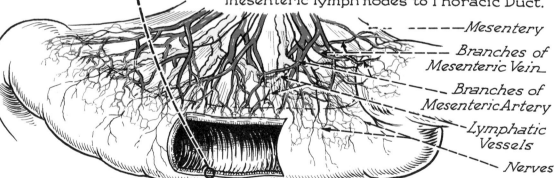

- - - - - Mesentery

Branches of Mesenteric Vein

Branches of Mesenteric Artery

Lymphatic Vessels

Nerves

Absorption is not just a process of simple diffusion of substances from areas of high to areas of low concentration. Movement of ions can take place against a concentration gradient. In other words — Absorption is often an active process involving the use of energy by the epithelial cells.

TRANSPORT of ABSORBED FOODSTUFFS

After absorption the NUTRIENTS are transported in:—

(a) <u>BLOOD</u> through
Mesenteric Veins
to
Portal Vein
to
LIVER— *{Many substances*
undergo further
changes;
some are stored;
some are passed on
via↓
Hepatic Veins → Inferior
Vena Cava →
Heart and
Systemic
Circulation

(b) <u>LYMPH</u> in
Lymphatic Vessels
to
THORACIC DUCT
and via a large vein
in the neck to
Systemic Circulation
↓
which then distributes
Food to ALL TISSUES
of the BODY

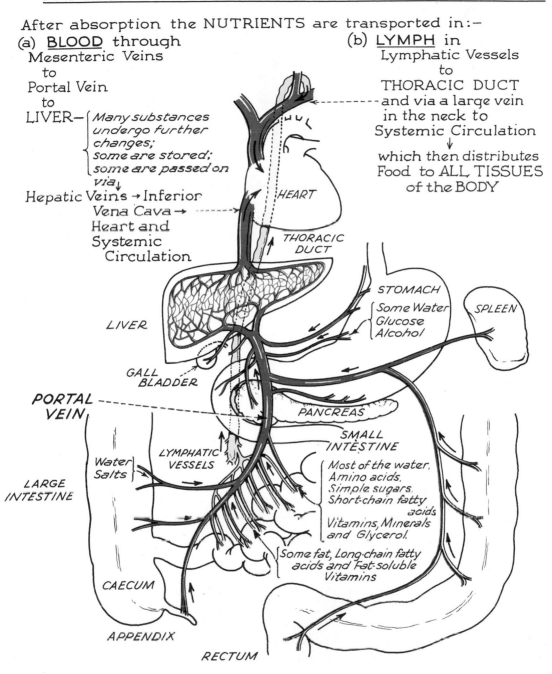

HEART

THORACIC DUCT

LIVER

STOMACH
{Some Water
Glucose
Alcohol

SPLEEN

GALL BLADDER

PORTAL VEIN

PANCREAS

SMALL INTESTINE

Water Salts

LYMPHATIC VESSELS

Most of the water,
Amino acids,
Simple sugars,
Short-chain fatty
acids
Vitamins, Minerals
and Glycerol.

Some fat, Long-chain fatty
acids and Fat-soluble
Vitamins

LARGE INTESTINE

CAECUM

APPENDIX

RECTUM

LARGE INTESTINE

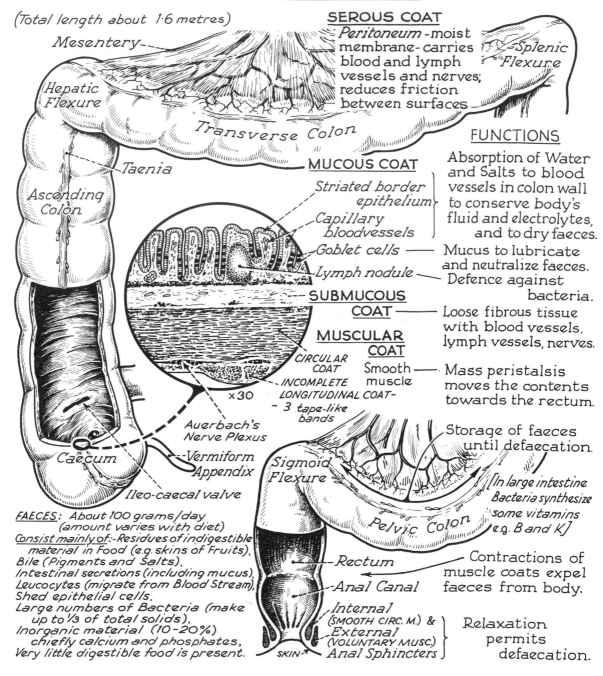

(Total length about 1·6 metres)

Mesentery

Hepatic Flexure

Transverse Colon

Taenia

Ascending Colon

Splenic Flexure

SEROUS COAT
Peritoneum - moist membrane - carries blood and lymph vessels and nerves; reduces friction between surfaces

MUCOUS COAT
Striated border epithelium

Capillary bloodvessels

Goblet cells

Lymph nodule

SUBMUCOUS COAT

MUSCULAR COAT
CIRCULAR COAT
INCOMPLETE LONGITUDINAL COAT - 3 tape-like bands

Smooth muscle

×30

Auerbach's Nerve Plexus

Caecum

Vermiform Appendix

Ileo-caecal valve

Sigmoid Flexure

Pelvic Colon

Rectum

Anal Canal

Internal (SMOOTH CIRC. M.) & External (VOLUNTARY MUSC.) Anal Sphincters

SKIN

FUNCTIONS

Absorption of Water and Salts to blood vessels in colon wall to conserve body's fluid and electrolytes, and to dry faeces.

Mucus to lubricate and neutralize faeces. Defence against bacteria.

Loose fibrous tissue with blood vessels, lymph vessels, nerves.

Mass peristalsis moves the contents towards the rectum.

Storage of faeces until defaecation.

[In large intestine Bacteria synthesize some vitamins e.g. B and K]

Contractions of muscle coats expel faeces from body.

Relaxation permits defaecation.

FAECES: About 100 grams/day (amount varies with diet)
Consist mainly of:- Residues of indigestible material in Food (e.g. skins of Fruits),
Bile (Pigments and Salts),
Intestinal secretions (including mucus),
Leucocytes (migrate from Blood Stream),
Shed epithelial cells,
Large numbers of Bacteria (make up to 1/3 of total solids),
Inorganic material (10-20%) chiefly calcium and phosphates,
Very little digestible food is present.

MOVEMENTS of the LARGE INTESTINE

The Ileo-caecal Valve opens and closes during digestion. Peristaltic waves sweep semi-fluid contents of Ileum through the relaxed valve. During its stay in the Large Intestine faecal matter is subjected to –

TWO TYPES OF MOVEMENT

"TONE WAVES" — — — — — — — — — — — — →
run forwards –
and backwards from
Cannon's Point to the
Ileum – (sometimes
mistakenly called
"Anti-peristalsis")
These waves "die out"

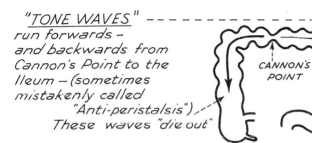

CANNON'S POINT

Ensure prolonged contact of contents with mucosa —and promote absorption of Water and Salt from faeces —
This type of movement is MYOGENIC

MASS PERISTALSIS
Strong waves — — — — — — — — — — — — →
at infrequent
intervals start
at upper end of
Ascending Colon

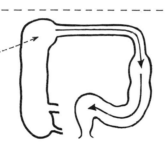

Empty Transverse Colon and sweep faeces into the Descending and Pelvic Colons and into Rectum —
—NEUROGENIC

[Reflex often initiated by passage of food into Stomach (Gastro-colic Reflex)]

EMPTYING
(DEFAECATION)

Complex
REFLEX act

Stimulus
Passage of
Faeces into
the Rectum
distends
wall

[
+
passage
through
anal
canal
]

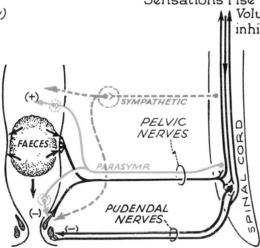

(+)

SYMPATHETIC

PELVIC NERVES

FAECES

PARASYMP.

SPINAL CORD

(−)

PUDENDAL NERVES

(−)

Sensations rise to level of consciousness →
Voluntary decision → Impulses to inhibit or permit reflex evacuation

Outgoing nerve messages →
Powerful peristaltic contractions of Descending Colon, Pelvic Colon and Rectum.

[Preceded by inspiratory descent of diaphragm and voluntary contraction of abdominal muscles to raise intra-abdominal pressure]

Contraction of Pelvic Floor muscles with Relaxation of Anal Sphincter
↓
Evacuation of Faeces

INNERVATION of the GUT WALL

The movements of the Alimentary Canal are carried out automatically and, for the most part, beneath the level of consciousness.

Like other "automatic" systems in the body the gut wall is supplied by nerves of the AUTONOMIC NERVOUS SYSTEM. This has TWO separate parts – SYMPATHETIC and PARASYMPATHETIC. They work together to achieve <u>balanced control</u> of events.

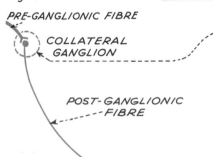

PRE-GANGLIONIC FIBRE

COLLATERAL GANGLION

POST-GANGLIONIC FIBRE

SYMPATHETIC nerves when stimulated release at their synapses a chemical substance, *ACETYLCHOLINE*, and at their endings in the muscle, *NORADRENALINE*. This *inhibits (–) i.e. causes relaxation* of the gut wall, because β receptors in the muscle are affected, and *stimulates (+) i.e. causes contracture or closure of the sphincters which have α receptors.*

Sympathetic post-ganglionic nerves terminate mainly in Auerbach's plexus except in the regions of the cardiac and anal sphincters.

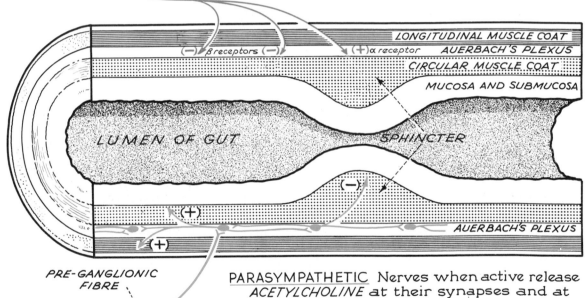

LONGITUDINAL MUSCLE COAT
AUERBACH'S PLEXUS
CIRCULAR MUSCLE COAT
MUCOSA AND SUBMUCOSA
(–) β receptors (–)   (+) α receptor
LUMEN OF GUT   SPHINCTER
(–)
(+)
(+)
AUERBACH'S PLEXUS

PRE-GANGLIONIC FIBRE

PARASYMPATHETIC Nerves when active release *ACETYLCHOLINE* at their synapses and at their endings in the muscle.
This *stimulates (+)* the gut wall to contract and *inhibits (–)*, or relaxes, the sphincters.

NERVOUS CONTROL of GUT MOVEMENTS

Movements in the wall of the Alimentary Canal are

either $\left\{\begin{array}{l}\text{(a) } MYOGENIC — \textit{a property of the smooth muscle.}\\ \text{(b) } NEUROGENIC — \textit{dependent on the Intrinsic Nerve Plexuses.}\end{array}\right.$

They can occur even after *Extrinsic* nerves to the tract have been cut. Normally, however, impulses travelling in these nerves of the SYMPATHETIC and PARASYMPATHETIC Systems, from the CONTROLLING CENTRES of the AUTONOMIC NERVOUS SYSTEM in the BRAIN, and SPINAL CORD, *influence* and *co-ordinate* events in the whole tract.

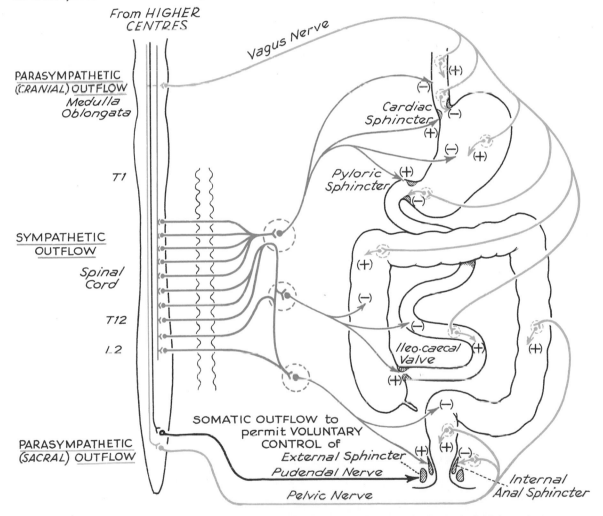

Parasympathetic promotes vegetative functions: Sympathetic inhibits them.

CHAPTER 4

TRANSPORT SYSTEM

The HEART, BLOOD VESSELS and BODY FLUIDS: HAEMOPOIETIC SYSTEM

CARDIOVASCULAR SYSTEM

The CIRCULATORY System

Chief TRANSPORT System of the body

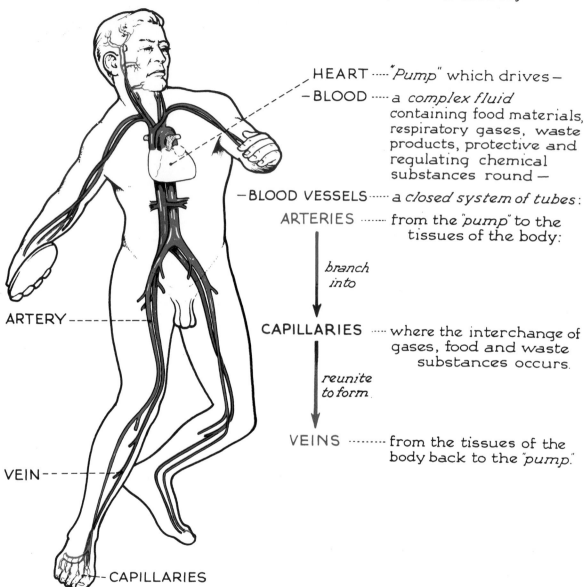

HEART ----- *"Pump"* which drives —

— BLOOD ----- a *complex fluid* containing food materials, respiratory gases, waste products, protective and regulating chemical substances round —

— BLOOD VESSELS ----- a *closed system of tubes:*

ARTERIES ------- from the *"pump"* to the tissues of the body:

branch into

CAPILLARIES ---- where the interchange of gases, food and waste substances occurs.

reunite to form

VEINS ------- from the tissues of the body back to the *"pump."*

ARTERY -----

VEIN -----

-- CAPILLARIES

GENERAL COURSE of the CIRCULATION

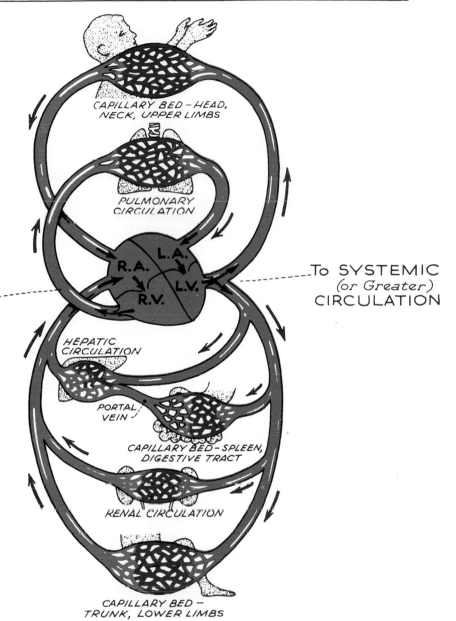

CAPILLARY BED — HEAD,
NECK, UPPER LIMBS

PULMONARY
CIRCULATION

R.A.   L.A.

L.V.

R.V.

To PULMONARY
(or Lesser)
CIRCULATION

To SYSTEMIC
(or Greater)
CIRCULATION

HEPATIC
CIRCULATION

PORTAL
VEIN

CAPILLARY BED — SPLEEN,
DIGESTIVE TRACT

RENAL CIRCULATION

CAPILLARY BED —
TRUNK, LOWER LIMBS

HEART

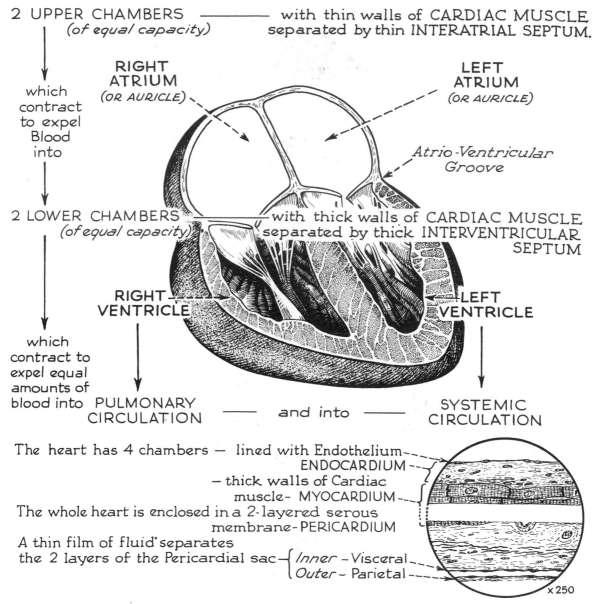

2 UPPER CHAMBERS ————— with thin walls of CARDIAC MUSCLE
(of equal capacity) separated by thin INTERATRIAL SEPTUM.

RIGHT ATRIUM *(OR AURICLE)*

LEFT ATRIUM *(OR AURICLE)*

Atrio-Ventricular Groove

which contract to expel Blood into

2 LOWER CHAMBERS *(of equal capacity)* with thick walls of CARDIAC MUSCLE separated by thick INTERVENTRICULAR SEPTUM

RIGHT VENTRICLE

LEFT VENTRICLE

which contract to expel equal amounts of blood into

PULMONARY CIRCULATION ——— and into ——— SYSTEMIC CIRCULATION

The heart has 4 chambers — lined with Endothelium — ENDOCARDIUM
— thick walls of Cardiac muscle- MYOCARDIUM
The whole heart is enclosed in a 2-layered serous membrane- PERICARDIUM
A thin film of fluid separates the 2 layers of the Pericardial sac — { *Inner* – Visceral
Outer – Parietal

x 250

The Cardiac muscle of the Atria is completely separated from the Cardiac muscle of the Ventricles by a ring of Fibrous Tissue — at the Atrio-Ventricular Groove.

HEART

This is a *diagrammatic* section through the heart.

<u>HEART VALVES</u> have a core of Fibrous Tissue ------ covered on both sides with *Endothelium*.

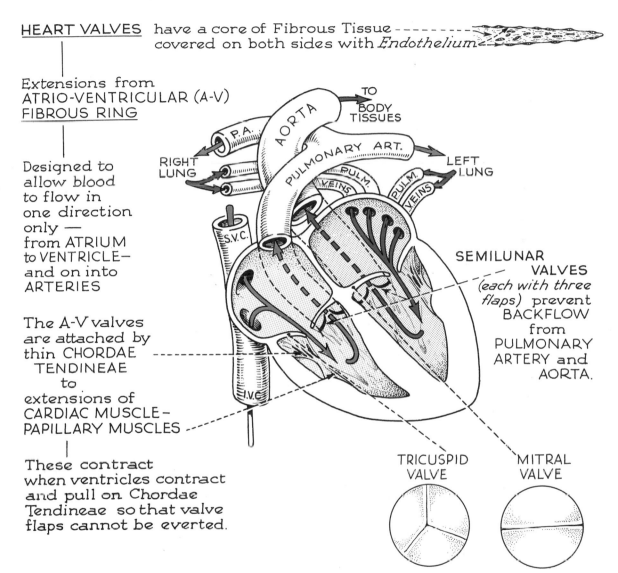

Extensions from
<u>ATRIO-VENTRICULAR (A-V)
FIBROUS RING</u>

Designed to
allow blood
to flow in
one direction
only —
from ATRIUM
to VENTRICLE—
and on into
ARTERIES

The A-V valves
are attached by
thin CHORDAE
TENDINEAE
to
extensions of
CARDIAC MUSCLE—
PAPILLARY MUSCLES

These contract
when ventricles contract
and pull on Chordae
Tendineae so that valve
flaps cannot be everted.

AORTA

P.A.

TO
BODY
TISSUES

RIGHT
LUNG

PULMONARY ART.

LEFT
LUNG

PULM.
VEINS

PULM.
VEINS

S.V.C.

I.V.C.

SEMILUNAR
VALVES
*(each with three
flaps)* prevent
BACKFLOW
from
PULMONARY
ARTERY and
AORTA.

TRICUSPID
VALVE

MITRAL
VALVE

*The Great Veins do not have valves guarding their entrance to the heart.
Thickening and contraction of the muscle around their mouths
prevent BACKFLOW of blood from heart.*

HEART

The human heart is really a *DOUBLE PUMP* — each quite separate from the other

RIGHT   *LEFT*

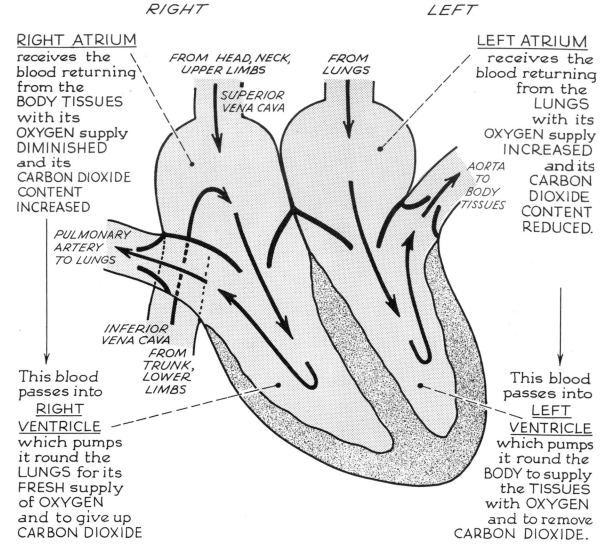

RIGHT ATRIUM
receives the blood returning from the BODY TISSUES with its OXYGEN supply DIMINISHED and its CARBON DIOXIDE CONTENT INCREASED

FROM HEAD, NECK, UPPER LIMBS

FROM LUNGS

SUPERIOR VENA CAVA

LEFT ATRIUM
receives the blood returning from the LUNGS with its OXYGEN supply INCREASED and its CARBON DIOXIDE CONTENT REDUCED.

AORTA TO BODY TISSUES

PULMONARY ARTERY TO LUNGS

INFERIOR VENA CAVA

FROM TRUNK, LOWER LIMBS

This blood passes into **RIGHT VENTRICLE**
which pumps it round the LUNGS for its FRESH supply of OXYGEN and to give up CARBON DIOXIDE

This blood passes into **LEFT VENTRICLE**
which pumps it round the BODY to supply the TISSUES with OXYGEN and to remove CARBON DIOXIDE.

This diagram simplifies the structure of the heart to make it easier to understand the function of its various parts.

CARDIAC CYCLE

Diagrammatic representation of the sequence of events in the heart during ONE heart beat.

DIASTOLE *[Period of Relaxation - i.e. when heart is resting]*

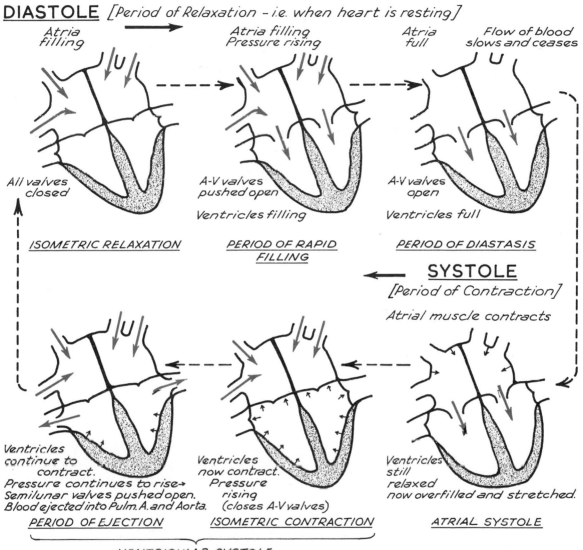

Atria filling

Atria filling
Pressure rising

Atria full

Flow of blood slows and ceases

All valves closed

A-V valves pushed open

Ventricles filling

A-V valves open

Ventricles full

ISOMETRIC RELAXATION

PERIOD OF RAPID FILLING

PERIOD OF DIASTASIS

SYSTOLE

[Period of Contraction]

Atrial muscle contracts

Ventricles continue to contract.
Pressure continues to rise→
Semilunar valves pushed open.
Blood ejected into Pulm. A. and Aorta.

Ventricles now contract.
Pressure rising
(closes A-V valves)

Ventricles still relaxed now overfilled and stretched.

PERIOD OF EJECTION

ISOMETRIC CONTRACTION

ATRIAL SYSTOLE

VENTRICULAR SYSTOLE

The total cycle of events takes about 0·8 second when heart is beating 75 times per minute.

HEART SOUNDS

During each CARDIAC CYCLE 2 HEART SOUNDS can be heard through a STETHOSCOPE applied to the CHEST WALL.

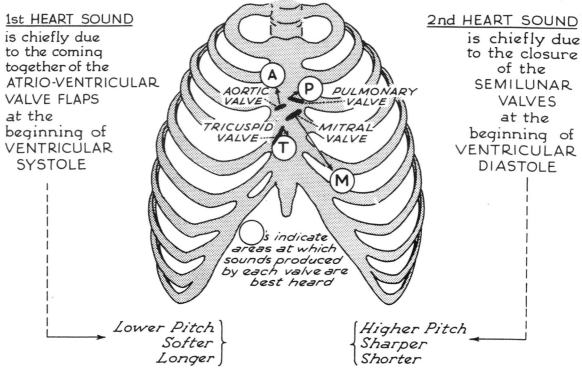

1st HEART SOUND
is chiefly due to the coming together of the ATRIO-VENTRICULAR VALVE FLAPS at the beginning of VENTRICULAR SYSTOLE

AORTIC VALVE

PULMONARY VALVE

TRICUSPID VALVE

MITRAL VALVE

○s indicate areas at which sounds produced by each valve are best heard

2nd HEART SOUND
is chiefly due to the closure of the SEMILUNAR VALVES at the beginning of VENTRICULAR DIASTOLE

Lower Pitch
Softer
Longer

Higher Pitch
Sharper
Shorter

The sounds may be represented phonetically:-

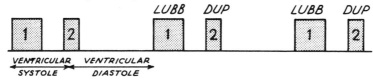

LUBB   DUP   LUBB   DUP

| 1 | 2 | 1 | 2 | 1 | 2 |

VENTRICULAR VENTRICULAR
SYSTOLE   DIASTOLE

They are repeated with every CARDIAC CYCLE i.e. about 70 times per minute in the average healthy adult.

If the valves have been damaged by disease additional murmurs can be heard as the blood flows forwards through narrowed valves or leaks backwards through incompetent valves.

Occasionally additional sounds are heard over normal healthy hearts.

ORIGIN and CONDUCTION of the HEART BEAT

The rhythmic contraction of the heart is called the HEART BEAT. The impulse to contract is generated in specialized NODAL TISSUE in the wall of the RIGHT ATRIUM.

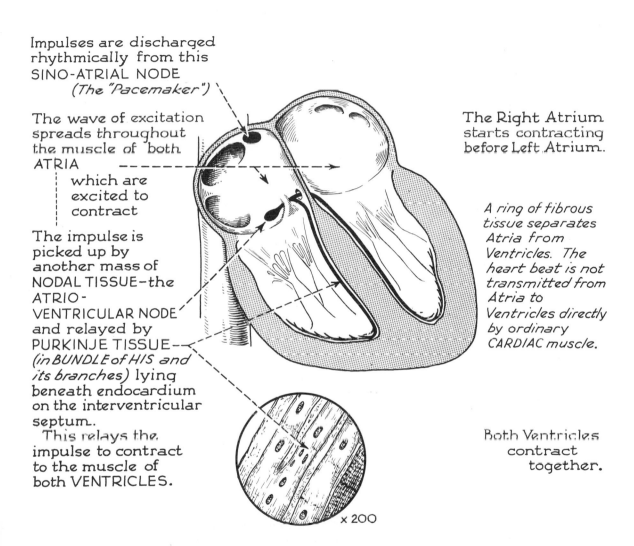

Impulses are discharged rhythmically from this SINO-ATRIAL NODE *(The "Pacemaker")*

The wave of excitation spreads throughout the muscle of both ATRIA

which are excited to contract

The impulse is picked up by another mass of NODAL TISSUE—the ATRIO-VENTRICULAR NODE and relayed by PURKINJE TISSUE—*(in BUNDLE of HIS and its branches)* lying beneath endocardium on the interventricular septum.
This relays the impulse to contract to the muscle of both VENTRICLES.

The Right Atrium starts contracting before Left Atrium.

A ring of fibrous tissue separates Atria from Ventricles. The heart beat is not transmitted from Atria to Ventricles directly by ordinary CARDIAC muscle.

Both Ventricles contract together.

x 200

The wave of excitation is accompanied by an electrical change which is followed within 0·02 seconds by contraction of the cardiac muscle.

ELECTROCARDIOGRAM

The wave of excitation spreading through the heart wall is accompanied by electrical changes. *(Like Nerve and Skeletal Muscle, active Cardiac Muscle is electrically negative relative to resting Cardiac Muscle ahead of the zone of excitation.)* The electrical currents produced are conducted to the surface of the body and can be picked up, amplified and recorded by a special instrument – the *ELECTROCARDIOGRAPH*.

The record obtained is the *ELECTROCARDIOGRAM : (E.C.G.)*

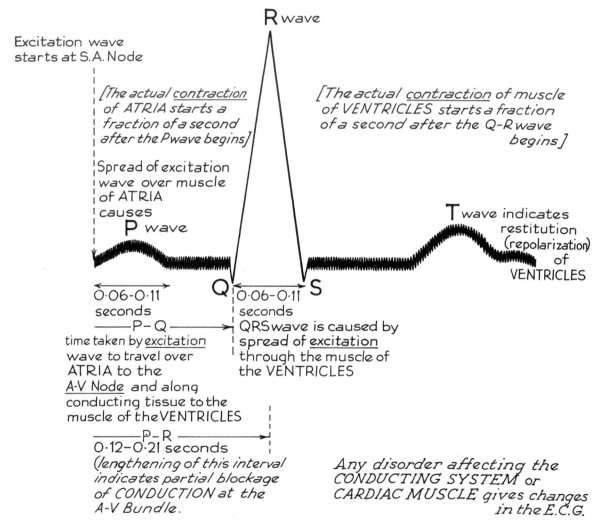

R wave

Excitation wave starts at S.A. Node

[The actual *contraction* of ATRIA starts a fraction of a second after the P wave begins]

[The actual *contraction* of muscle of VENTRICLES starts a fraction of a second after the Q–R wave begins]

Spread of excitation wave over muscle of ATRIA causes

P wave

T wave indicates restitution (repolarization) of VENTRICLES

Q   S

0·06–0·11 seconds

0·06–0·11 seconds

————— P–Q —————

QRS wave is caused by spread of *excitation* through the muscle of the VENTRICLES

time taken by *excitation* wave to travel over ATRIA to the A-V Node and along conducting tissue to the muscle of the VENTRICLES

————— P–R —————

0·12–0·21 seconds
(lengthening of this interval indicates partial blockage of CONDUCTION at the A-V Bundle.

Any disorder affecting the CONDUCTING SYSTEM or CARDIAC MUSCLE gives changes in the E.C.G.

NERVOUS REGULATION of ACTION of HEART

Although the heart initiates its own impulse to contraction its activity is finely adjusted, to meet the body's constantly changing needs, by nervous impulses discharged from CONTROLLING CENTRES in the BRAIN and SPINAL CORD along PARASYMPATHETIC and SYMPATHETIC OUTFLOWS.

ACTION of PARASYMPATHETIC —

Continuous stream of impulses to the PACEMAKER tends to restrain (–) Heart's Action:- SLOWS HEART RATE, CONDUCTION at A.V. NODE DELAYED, FORCE of CONTRACTION DECREASED, EXCITABILITY DECREASED.

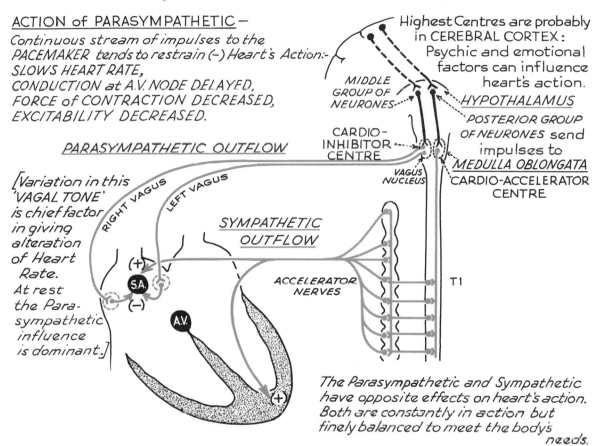

Highest Centres are probably in CEREBRAL CORTEX: Psychic and emotional factors can influence heart's action.

MIDDLE GROUP OF NEURONES

HYPOTHALAMUS
POSTERIOR GROUP OF NEURONES send impulses to MEDULLA OBLONGATA CARDIO-ACCELERATOR CENTRE

CARDIO-INHIBITOR CENTRE

VAGUS NUCLEUS

PARASYMPATHETIC OUTFLOW

RIGHT VAGUS   LEFT VAGUS

SYMPATHETIC OUTFLOW

[Variation in this 'VAGAL TONE' is chief factor in giving alteration of Heart Rate. At rest the Parasympathetic influence is dominant.]

ACCELERATOR NERVES

T1

(+)   ∨   S.A.   (–)

A.V.

(+)

The Parasympathetic and Sympathetic have opposite effects on heart's action. Both are constantly in action but finely balanced to meet the body's needs.

ACTION of SYMPATHETIC —

Stimulation of the sympathetic accelerates (+) Heart's Action:- RATE of CONTRACTION INCREASED, CONDUCTIVITY INCREASED, FORCE of CONTRACTION INCREASED, EXCITABILITY INCREASED.

Stimulation of Symp. Inhibition of Para.	→Acceleration	of Heart's Action
Stimulation of Para. Inhibition of Symp.	→Inhibition	

[The Sympathetic influence is dominant e.g. in stress, exercise, excessive heat, and other conditions requiring rapid blood flow.]

CARDIAC REFLEXES

There are INGOING *(SENSORY)* fibres travelling in both PARASYMPATHETIC and SYMPATHETIC nerves which convey information to the CENTRES about events taking place in the heart.

These afferent messages do not normally reach consciousness. They are important as the AFFERENT pathways in CARDIAC REFLEXES by means of which heart's action is adjusted to body's needs.

AUTONOMIC CENTRES IN MEDULLA OBLONGATA

BAINBRIDGE REFLEX

AFFERENTS in *Cranial X (Vagus)*

RECEPTORS *Stretch receptors near mouths of GREAT VEINS stimulated*

Stretching of Receptors by rise in pressure in Right Atrium.

STIMULUS *Increased return of venous blood to Right Atrium*

EFFERENTS *Decreased number of impulses sent out along* PARASYMPATHETIC OUTFLOW *to S.A.node*

Increased number of impulses sent out along SYMPATHETIC OUTFLOW *to S.A.node and to* VENTRICULAR MUSCLE

↑rate and ↑force of contraction
↓
↑cardiac output

This reflex augments Starling's law of the heart.

N.B. The opposite effects follow a fall in venous return.

It is an important adaptive mechanism whereby heart rate and force of contraction are reflexly adjusted to match the quantity of venous blood returning to the heart.

CARDIAC REFLEXES

One of the most important of these is the
BARORECEPTOR REFLEX for reflex control
of arterial blood pressure.

AFFERENTS

*in IX and X
Cranial nerves*

RECEPTORS
*Stretch (or Baro.)
receptors in* CAROTID
SINUS
ARCH of
AORTA
*and probably
in* MYOCARDIUM

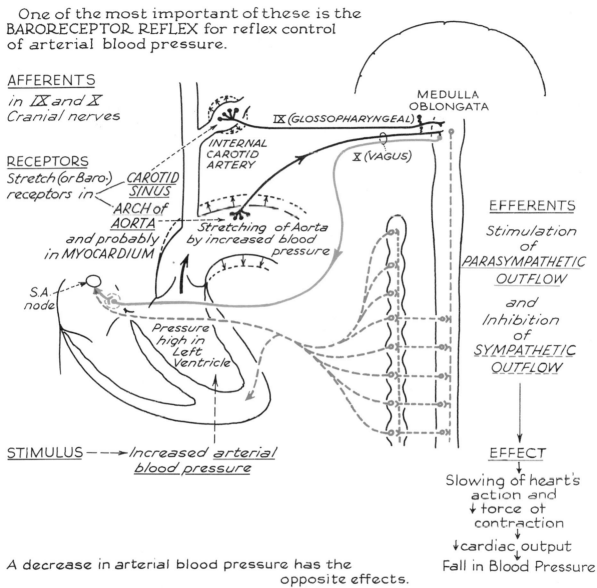

MEDULLA
OBLONGATA

IX (GLOSSOPHARYNGEAL)

INTERNAL
CAROTID
ARTERY

X (VAGUS)

*Stretching of Aorta
by increased blood
pressure*

EFFERENTS

*Stimulation
of*
PARASYMPATHETIC
OUTFLOW

*and
Inhibition
of*
SYMPATHETIC
OUTFLOW

S.A.
node

*Pressure
high in
Left
Ventricle*

STIMULUS ---> *Increased arterial
blood pressure*

EFFECT

Slowing of heart's
action and
↓ force of
contraction

↓ cardiac output
Fall in Blood Pressure

A decrease in arterial blood pressure has the
opposite effects.

*Important Adaptive Mechanism — As Arterial Blood Pressure is raised
by ejection of blood from Ventricle in Systole this reflex is brought
into play to adjust force and frequency with which heart is ejecting
blood into the arterial system. It is important in keeping B.P. stable.*

CARDIAC OUTPUT

The Volume of Blood expelled from the heart can be measured as follows :-

A semi-rigid tube is inserted into a VEIN in the arm and passed along into the RIGHT ATRIUM of the heart — to obtain a sample of MIXED VENOUS BLOOD — which has given up some of its oxygen to the tissues.
The OXYGEN content is analysed.

100ml VENOUS Blood hold 14ml OXYGEN.

The AMOUNT of OXYGEN taken up by the lungs per minute is measured by a SPIROMETER.

250 ml OXYGEN are removed from the lungs by blood each minute.

A needle is inserted into an ARTERY in the leg and a sample of ARTERIAL BLOOD — which has received its fresh oxygen supply in the lungs — is obtained.
The OXYGEN content is analysed.

100ml ARTERIAL Blood hold 19ml OXYGEN.

Each 100ml blood gains 5ml OXYGEN as it passes through the lungs. The blood in the lungs takes up 250 ml OXYGEN from the atmosphere per minute.

Therefore there must be $\left(\frac{250}{5} \times 100\right)$ ml *[i.e. 5,000 ml] of blood leaving the Right Ventricle and passing through the Lungs to the Left Atrium per minute to take up this* 250ml OXYGEN.

(After WISHART)

The same volume of blood must leave the Left Ventricle and enter the Aorta in the same time otherwise blood would soon be dammed back in the lungs. i.e. If heart contracts 72 times per minute STROKE VOLUME = $\frac{5000}{72} \doteqdot$ *70 ml per beat for each Ventricle.*

Cardiac Output can be increased many times *(up to 30 litres per min)* in Exercise —
— partly by increase in Heart Rate; partly by increase in
Stroke Volume.

90

BLOOD VESSELS

The system of tubes through which the heart pumps blood.

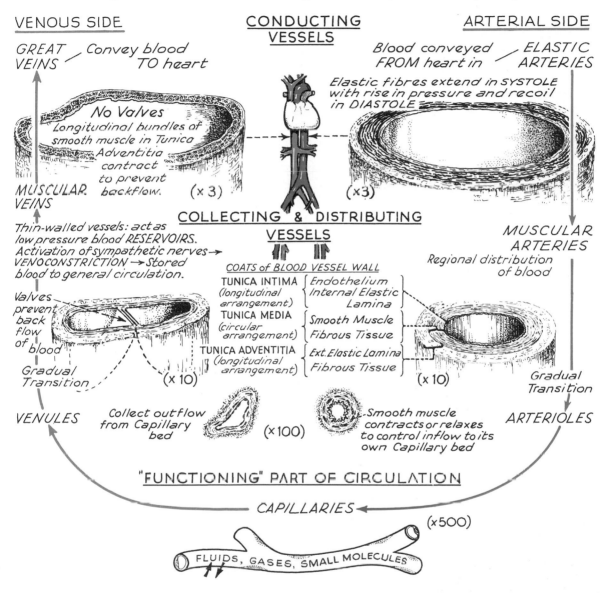

VENOUS SIDE

GREAT VEINS — Convey blood TO heart

No Valves
Longitudinal bundles of smooth muscle in Tunica Adventitia contract to prevent backflow. (x 3)

MUSCULAR VEINS

Thin-walled vessels: act as low pressure blood RESERVOIRS. Activation of sympathetic nerves → VENOCONSTRICTION → Stored blood to general circulation.

Valves prevent back flow of blood

Gradual Transition (x 10)

VENULES

Collect outflow from Capillary bed (x 100)

CONDUCTING VESSELS

COLLECTING & DISTRIBUTING VESSELS

COATS of BLOOD VESSEL WALL

TUNICA INTIMA (longitudinal arrangement) — Endothelium / Internal Elastic Lamina

TUNICA MEDIA (circular arrangement) — Smooth Muscle / Fibrous Tissue

TUNICA ADVENTITIA (longitudinal arrangement) — Ext. Elastic Lamina / Fibrous Tissue

ARTERIAL SIDE

Blood conveyed FROM heart in — ELASTIC ARTERIES

Elastic fibres extend in SYSTOLE with rise in pressure and recoil in DIASTOLE (x3)

MUSCULAR ARTERIES
Regional distribution of blood

Gradual Transition (x 10)

Smooth muscle contracts or relaxes to control inflow to its own Capillary bed

ARTERIOLES

"FUNCTIONING" PART OF CIRCULATION

CAPILLARIES ◄— (x500)

FLUIDS, GASES, SMALL MOLECULES

Only from CAPILLARIES can Blood give up food and oxygen to tissues; and receive waste products and carbon dioxide from tissues.

BLOOD PRESSURE

Each Ventricle at each heart beat ejects forcibly about 70 ml. of Blood into the Blood Vessels. All of this blood cannot pass through arterioles into capillaries and veins during the heart's contraction. This means that roughly 5/8 of the <u>CARDIAC OUTPUT</u> at each heart beat has to be stored during systole and passed on during diastole.

<u>CONDUCTING ARTERIES</u> are always more or less stretched.

<u>PERIPHERAL RESISTANCE</u> is offered to the passage of blood from arterial to venous side of the system chiefly by partial constriction ("Tone") of smooth muscle in walls of Arterioles. *(The calibre is regulated by action of Parasympathetic and Sympathetic Nervous System — [see pages 95-97]*

These factors are largely responsible for the considerable pressure of the blood in the Arterial System. The pressure is highest at the height of the heart's contraction, i.e. <u>SYSTOLIC BLOOD PRESSURE</u>, and lowest when the heart is relaxing, i.e. <u>DIASTOLIC BLOOD PRESSURE</u>.

The pressure is LOWEST as blood drains into RIGHT ATRIUM at end of DIASTOLE.

The pressure is HIGHEST as blood leaves the LEFT VENTRICLE at end of SYSTOLE.

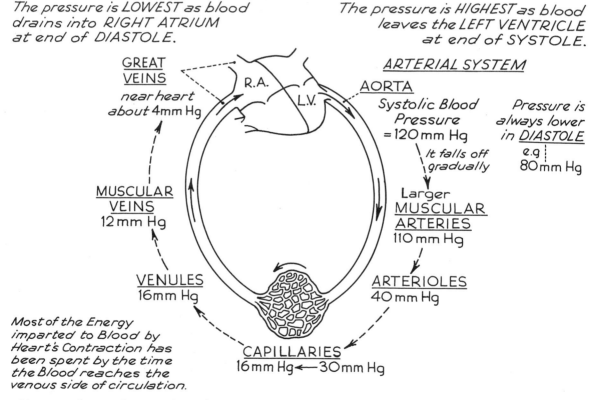

<u>GREAT VEINS</u>
near heart
about 4mm Hg

R.A.

L.V.

<u>ARTERIAL SYSTEM</u>

<u>AORTA</u>
Systolic Blood Pressure
=120 mm Hg

Pressure is always lower in DIASTOLE e.g 80 mm Hg

It falls off gradually

Larger <u>MUSCULAR ARTERIES</u>
110 mm Hg

<u>MUSCULAR VEINS</u>
12 mm Hg

<u>VENULES</u>
16mm Hg

<u>ARTERIOLES</u>
40 mm Hg

Most of the Energy imparted to Blood by Heart's Contraction has been spent by the time the Blood reaches the venous side of circulation.

<u>CAPILLARIES</u>
16mm Hg ← 30mm Hg

NOTE :- *Any alteration in the <u>TOTAL AMOUNT</u> or <u>VISCOSITY</u> of Blood will also affect BLOOD PRESSURE.*

MEASUREMENT of ARTERIAL BLOOD PRESSURE

The Arterial Blood Pressure is measured in man by means of a _SPHYGMOMANOMETER_.

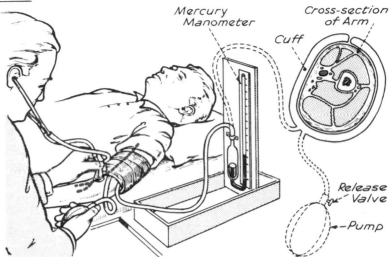

Mercury Manometer

Cross-section of Arm

Cuff

This consists of a RUBBER BAG (_covered with a cloth envelope_) which is wrapped round the UPPER ARM over the BRACHIAL ARTERY.

One tube connects the inside of the bag with a MANOMETER containing MERCURY.

Another tube connects the inside of the bag to a hand operated PUMP with a release VALVE.

Release Valve

Pump

METHOD

Air is pumped into the rubber bag till pressure in cuff is greater than pressure in artery even during heart's systole —
— Artery is then closed down during SYSTOLE and DIASTOLE
[At same time air is pushing up mercury column in manometer.]

By releasing valve on pump the pressure in cuff is gradually reduced till maximum pressure in artery just overcomes pressure in cuff — Some blood begins to spurt through during SYSTOLE — Artery still closed during DIASTOLE

At this point FAINT rhythmical TAPPING SOUNDS begin to be heard through STETHOSCOPE. The height of mercury in millimetres is taken as the SYSTOLIC Blood Pressure (e.g. 120 mm Hg)

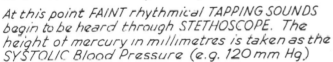

Pressure in cuff is reduced still further till it is just less than the lowest pressure in artery towards the end of diastole (i.e. just before next heart beat) —
— Blood flow is unimpeded during SYSTOLE and DIASTOLE
The sounds stop. The height of mercury in the manometer at this point is taken as the DIASTOLIC Blood Pressure (e.g. about 80mm Hg)

These values differ with SEX, AGE, EXERCISE, SLEEP, etc.

ELASTIC ARTERIES

The large <u>CONDUCTING ARTERIES</u> near the <u>HEART</u> are
<u>ELASTIC ARTERIES</u>

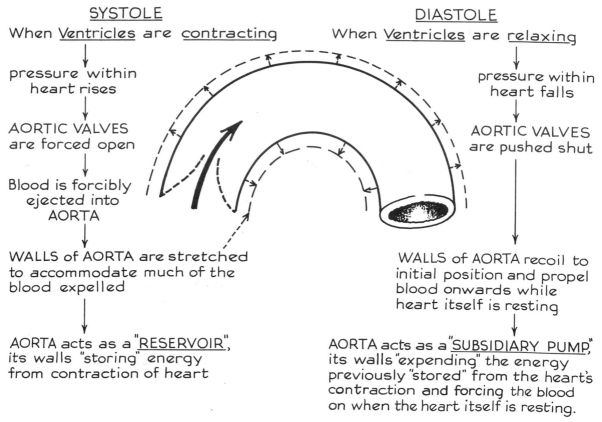

<u>SYSTOLE</u>	<u>DIASTOLE</u>
When <u>Ventricles</u> are <u>contracting</u>	When <u>Ventricles</u> are <u>relaxing</u>
pressure within heart rises	pressure within heart falls
AORTIC VALVES are forced open	AORTIC VALVES are pushed shut
Blood is forcibly ejected into AORTA	
WALLS of AORTA are stretched to accommodate much of the blood expelled	WALLS of AORTA recoil to initial position and propel blood onwards while heart itself is resting
AORTA acts as a "<u>RESERVOIR</u>", its walls "storing" energy from contraction of heart	AORTA acts as a "<u>SUBSIDIARY PUMP</u>," its walls "expending" the energy previously "stored" from the heart's contraction and forcing the blood on when the heart itself is resting.

As blood is pumped from the heart during systole, this distension and increase in pressure which starts in the aorta passes along the whole arterial system as a wave — the *pulse wave*.

The expansion and subsequent relaxation of the wall of the radial artery can be felt as "The PULSE" at the wrist.

A great increase in Blood Pressure can result if these walls lose some of their elasticity with age or disease and can no longer stretch readily to accommodate so much of the heart's output during systole: nor recoil so far in diastole.
The systolic and diastolic values may both be higher.

94

NERVOUS REGULATION of ARTERIOLES

The Smooth Muscle in *Arterioles* can contract or relax to alter calibre of the vessels, and thus adjust distribution of blood to meet the constantly changing needs of different parts of the body and also to maintain the normal blood pressure.

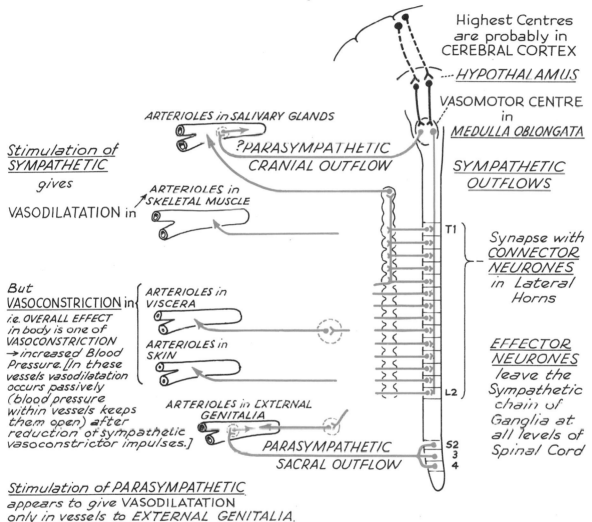

Highest Centres are probably in CEREBRAL CORTEX

----HYPOTHALAMUS

VASOMOTOR CENTRE in MEDULLA OBLONGATA

ARTERIOLES in SALIVARY GLANDS

?PARASYMPATHETIC CRANIAL OUTFLOW

Stimulation of SYMPATHETIC *gives*

VASODILATATION in

ARTERIOLES in SKELETAL MUSCLE

SYMPATHETIC OUTFLOWS

T1

Synapse with CONNECTOR NEURONES in Lateral Horns

But VASOCONSTRICTION in {

ARTERIOLES in VISCERA

ARTERIOLES in SKIN

i.e. OVERALL EFFECT in body is one of VASOCONSTRICTION →increased Blood Pressure. [In these vessels vasodilatation occurs passively (blood pressure within vessels keeps them open) after reduction of sympathetic vasoconstrictor impulses.]

ARTERIOLES in EXTERNAL GENITALIA

L2

EFFECTOR NEURONES leave the Sympathetic chain of Ganglia at all levels of Spinal Cord

PARASYMPATHETIC SACRAL OUTFLOW

S2
3
4

Stimulation of PARASYMPATHETIC
appears to give VASODILATATION
only in vessels to EXTERNAL GENITALIA.

VEINS have a sympathetic nerve supply. When activated these cause generalized venoconstriction to mobilize the blood held in "venous reservoirs".

REFLEX and CHEMICAL REGULATION of ARTERIOLAR TONE – I

Afferent impulses are constantly reaching the VASOMOTOR CENTRE in the Medulla Oblongata in ingoing nerves from all parts of the body. These form the AFFERENT PATHWAYS for VASOMOTOR REFLEXES.

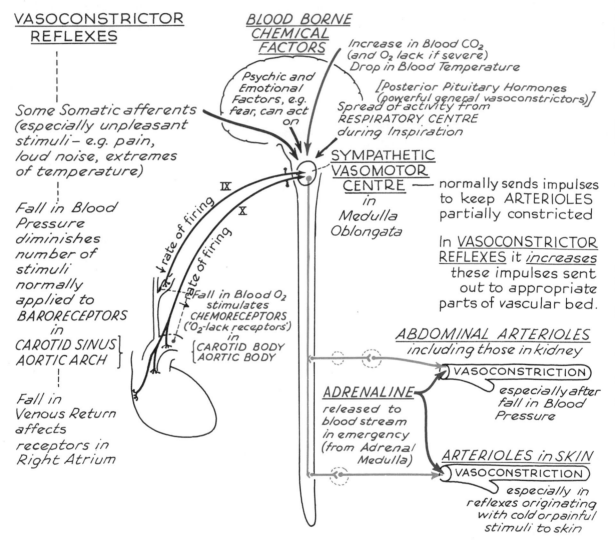

VASOCONSTRICTOR REFLEXES

Some Somatic afferents (especially unpleasant stimuli – e.g. pain, loud noise, extremes of temperature)

Fall in Blood Pressure diminishes number of stimuli normally applied to BARORECEPTORS in CAROTID SINUS] AORTIC ARCH }

Fall in Venous Return affects receptors in Right Atrium

BLOOD BORNE CHEMICAL FACTORS

Increase in Blood CO_2 (and O_2 lack if severe)
Drop in Blood Temperature

[Posterior Pituitary Hormones (powerful general vasoconstrictors)]

Spread of activity from RESPIRATORY CENTRE during Inspiration

Psychic and Emotional Factors, e.g. fear, can act on

IX ↑ rate of firing

X ↑ rate of firing

↑ rate of firing

Fall in Blood O_2 stimulates CHEMORECEPTORS ('O_2-lack receptors') in { CAROTID BODY AORTIC BODY }

SYMPATHETIC VASOMOTOR CENTRE — in Medulla Oblongata

normally sends impulses to keep ARTERIOLES partially constricted

In VASOCONSTRICTOR REFLEXES it *increases* these impulses sent out to appropriate parts of vascular bed.

ABDOMINAL ARTERIOLES including those in kidney

(VASOCONSTRICTION)
especially after fall in Blood Pressure

ADRENALINE released to blood stream in emergency (from Adrenal Medulla)

ARTERIOLES in SKIN

(VASOCONSTRICTION)
especially in reflexes originating with cold or painful stimuli to skin

VASOCONSTRICTION, especially of abdominal vessels, gives rise in Blood Pressure.

REFLEX and CHEMICAL REGULATION of ARTERIOLAR TONE – II

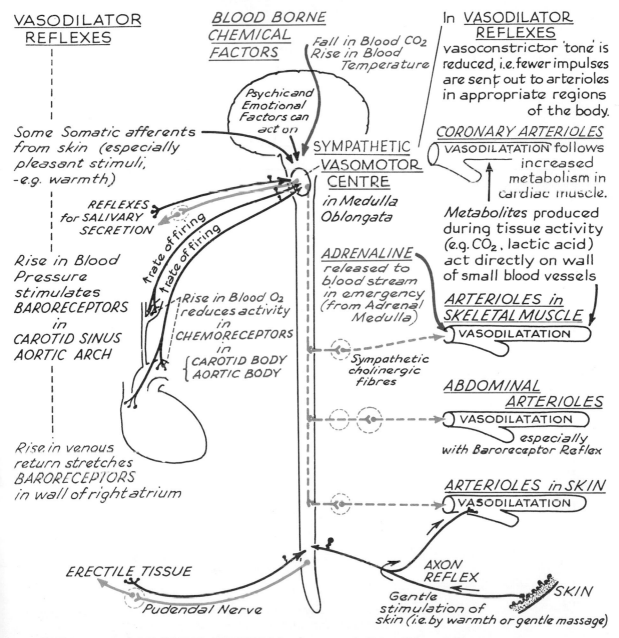

VASODILATOR REFLEXES

BLOOD BORNE CHEMICAL FACTORS

Fall in Blood CO_2 Rise in Blood Temperature

In VASODILATOR REFLEXES vasoconstrictor 'tone' is reduced, i.e. fewer impulses are sent out to arterioles in appropriate regions of the body.

Psychic and Emotional Factors can act on

Some Somatic afferents from skin (especially pleasant stimuli, -e.g. warmth)

SYMPATHETIC VASOMOTOR CENTRE in Medulla Oblongata

CORONARY ARTERIOLES
VASODILATATION follows increased metabolism in cardiac muscle.

REFLEXES for SALIVARY SECRETION

↑rate of firing

↑rate of firing

ADRENALINE released to blood stream in emergency (from Adrenal Medulla)

Metabolites produced during tissue activity (e.g. CO_2, lactic acid) act directly on wall of small blood vessels

Rise in Blood Pressure stimulates BARORECEPTORS in CAROTID SINUS AORTIC ARCH

Rise in Blood O_2 reduces activity in CHEMORECEPTORS in CAROTID BODY AORTIC BODY

ARTERIOLES in SKELETAL MUSCLE
VASODILATATION

Sympathetic cholinergic fibres

ABDOMINAL ARTERIOLES
VASODILATATION
especially with Baroreceptor Reflex

Rise in venous return stretches BARORECEPTORS in wall of right atrium

ARTERIOLES in SKIN
VASODILATATION

ERECTILE TISSUE

AXON REFLEX

SKIN

Pudendal Nerve

Gentle stimulation of skin (i.e. by warmth or gentle massage)

Widespread VASODILATATION gives a fall in Blood Pressure.

CAPILLARIES

Blood is distributed by *Arterioles* to *Capillaries*. Through the semi-permeable walls of *Capillaries* the sole purpose of the Circulating System is achieved.

FORCES determining EXCHANGE of WATER and ELECTROLYTES between BLOOD ——— and ——— TISSUE FLUIDS

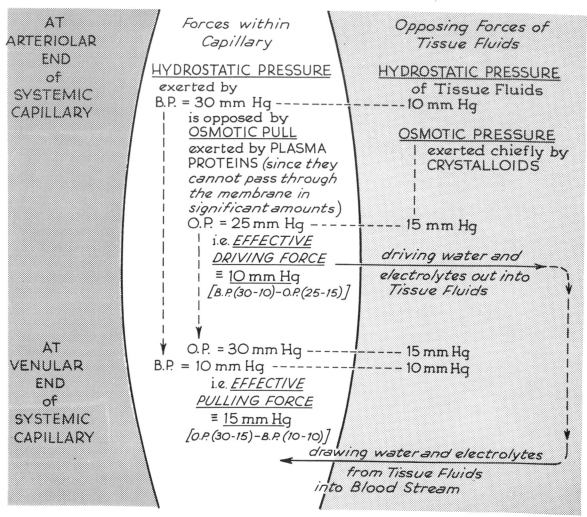

AT ARTERIOLAR END of SYSTEMIC CAPILLARY

Forces within Capillary

HYDROSTATIC PRESSURE
exerted by
B.P. = 30 mm Hg
is opposed by
OSMOTIC PULL
exerted by PLASMA PROTEINS (*since they cannot pass through the membrane in significant amounts*)
O.P. = 25 mm Hg
i.e. *EFFECTIVE DRIVING FORCE*
≡ 10 mm Hg
[*B.P.(30-10)-O.P.(25-15)*]

Opposing Forces of Tissue Fluids

HYDROSTATIC PRESSURE
of Tissue Fluids
10 mm Hg

OSMOTIC PRESSURE
exerted chiefly by
CRYSTALLOIDS
15 mm Hg

driving water and electrolytes out into Tissue Fluids

AT VENULAR END of SYSTEMIC CAPILLARY

O.P. = 30 mm Hg ----------- 15 mm Hg
B.P. = 10 mm Hg ----------- 10 mm Hg
i.e. *EFFECTIVE PULLING FORCE*
≡ 15 mm Hg
[*O.P.(30-15)-B.P.(10-10)*]

drawing water and electrolytes from Tissue Fluids into Blood Stream

These exchanges in Systemic Capillaries result in a continuous turn-over and renewal of TISSUE FLUIDS.

VEINS: VENOUS RETURN

Capillaries unite to form *Veins* which *convey blood back to the heart.* By the time blood reaches the veins much of the force imparted to it by the heart's contraction has been spent.

<u>VENOUS RETURN</u> to the heart depends on various factors :-

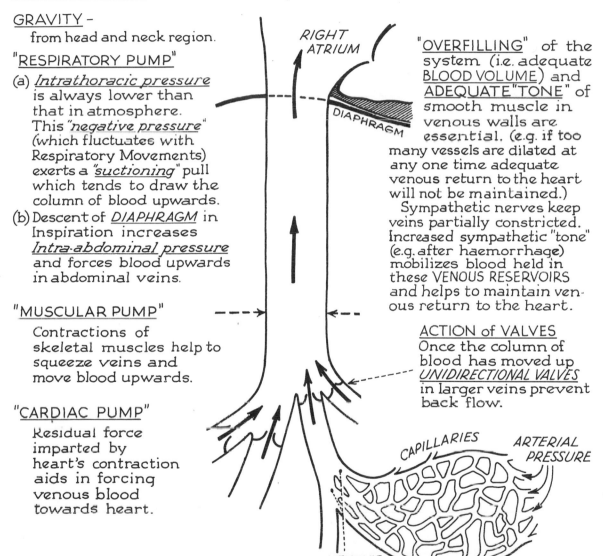

<u>GRAVITY</u> –
from head and neck region.

<u>"RESPIRATORY PUMP"</u>

(a) *Intrathoracic pressure* is always lower than that in atmosphere. This *"negative pressure"* (which fluctuates with Respiratory Movements) exerts a *"suctioning"* pull which tends to draw the column of blood upwards.

(b) Descent of *DIAPHRAGM* in Inspiration increases *Intra-abdominal pressure* and forces blood upwards in abdominal veins.

<u>"MUSCULAR PUMP"</u>

Contractions of skeletal muscles help to squeeze veins and move blood upwards.

<u>"CARDIAC PUMP"</u>

Residual force imparted by heart's contraction aids in forcing venous blood towards heart.

RIGHT ATRIUM

DIAPHRAGM

<u>"OVERFILLING"</u> of the system (i.e. adequate <u>BLOOD VOLUME</u>) and <u>ADEQUATE "TONE"</u> of smooth muscle in venous walls are essential. (e.g. if too many vessels are dilated at any one time adequate venous return to the heart will not be maintained.)

Sympathetic nerves keep veins partially constricted. Increased sympathetic "tone" (e.g. after haemorrhage) mobilizes blood held in these VENOUS RESERVOIRS and helps to maintain venous return to the heart.

<u>ACTION of VALVES</u>
Once the column of blood has moved up <u>UNIDIRECTIONAL VALVES</u> in larger veins prevent back flow.

CAPILLARIES   ARTERIAL PRESSURE

VENULES

BLOOD FLOW

The rate of blood flow varies in different parts of the vascular system. It is rapid in large vessels; slower in small vessels. *[The larger the total cross-sectional area represented the slower the rate of flow.]*

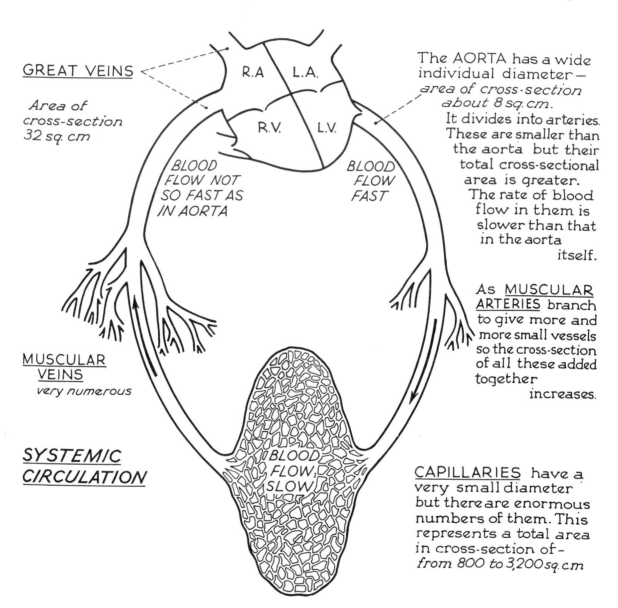

GREAT VEINS

Area of cross-section 32 sq. cm

R.A   L.A.

R.V.   L.V.

BLOOD FLOW NOT SO FAST AS IN AORTA

BLOOD FLOW FAST

The AORTA has a wide individual diameter — *area of cross-section about 8 sq. cm.* It divides into arteries. These are smaller than the aorta but their total cross-sectional area is greater. The rate of blood flow in them is slower than that in the aorta itself.

As MUSCULAR ARTERIES branch to give more and more small vessels so the cross-section of all these added together increases.

MUSCULAR VEINS
very numerous

SYSTEMIC CIRCULATION

BLOOD FLOW SLOW

CAPILLARIES have a very small diameter but there are enormous numbers of them. This represents a total area in cross-section of – *from 800 to 3,200 sq. cm*

100

PULMONARY CIRCULATION

BLOOD FLOW is faster in <u>PULMONARY CAPILLARIES</u> — Total cross-sectional area is smaller than that of *Systemic Capillaries*. Capillary pressure is less. This means that the osmotic "pull" of the plasma proteins (25 mm Hg) exceeds the "driving force" of the B.P. (5-10 mm Hg) along the whole length of the pulmonary capillary and fluid, therefore, cannot pass out into the alveoli.

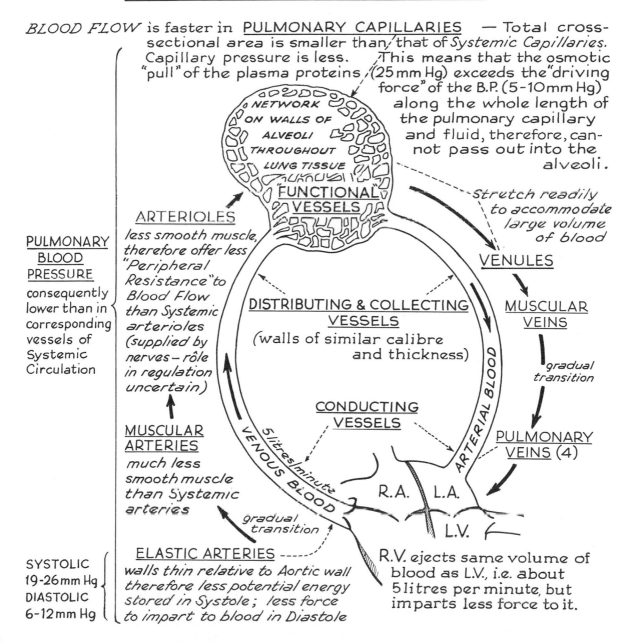

NETWORK ON WALLS OF ALVEOLI THROUGHOUT LUNG TISSUE
FUNCTIONAL VESSELS

Stretch readily to accommodate large volume of blood

<u>ARTERIOLES</u>
less smooth muscle, therefore offer less "Peripheral Resistance" to Blood Flow than Systemic arterioles (supplied by nerves – rôle in regulation uncertain)

<u>PULMONARY BLOOD PRESSURE</u>
consequently lower than in corresponding vessels of Systemic Circulation

DISTRIBUTING & COLLECTING VESSELS
(walls of similar calibre and thickness)

<u>VENULES</u>

<u>MUSCULAR VEINS</u>

gradual transition

ARTERIAL BLOOD

5 litres/minute VENOUS BLOOD

<u>CONDUCTING VESSELS</u>

<u>PULMONARY VEINS (4)</u>

<u>MUSCULAR ARTERIES</u>
much less smooth muscle than Systemic arteries

R.A.   L.A.

L.V.

gradual transition

SYSTOLIC 19-26 mm Hg
DIASTOLIC 6-12 mm Hg

<u>ELASTIC ARTERIES</u>
walls thin relative to Aortic wall therefore less potential energy stored in Systole; less force to impart to blood in Diastole

R.V. ejects same volume of blood as L.V., i.e. about 5 litres per minute, but imparts less force to it.

A great dilatation of Pulmonary vessels occurs in exercise.

DISTRIBUTION of WATER and ELECTROLYTES in BODY FLUIDS

Water makes up about 70% of Adult Human Body —
— i.e about 46 litres in 70 kg man.

Some lies outside the cells.
EXTRACELLULAR — 16 litres

A large part is within the cells.
INTRACELLULAR — 30 litres

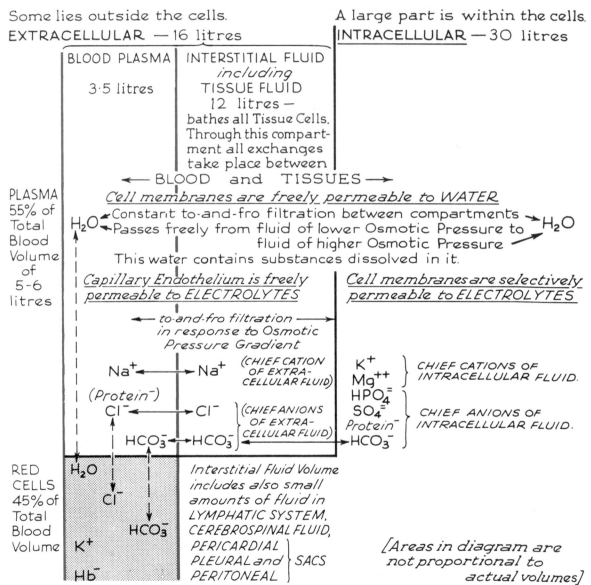

BLOOD PLASMA	INTERSTITIAL FLUID
3·5 litres	*including* TISSUE FLUID 12 litres — bathes all Tissue Cells. Through this compartment all exchanges take place between

PLASMA 55% of Total Blood Volume of 5-6 litres

← BLOOD and TISSUES →

Cell membranes are freely permeable to WATER

H_2O ← Constant to-and-fro filtration between compartments → H_2O
← Passes freely from fluid of lower Osmotic Pressure to fluid of higher Osmotic Pressure

This water contains substances dissolved in it.

Capillary Endothelium is freely permeable to ELECTROLYTES

Cell membranes are selectively permeable to ELECTROLYTES

← to-and-fro filtration → in response to Osmotic Pressure Gradient

Na^+ ← → Na^+ (CHIEF CATION OF EXTRA-CELLULAR FLUID)

K^+
Mg^{++} } CHIEF CATIONS OF INTRACELLULAR FLUID.

(Protein⁻)
Cl^- ← → Cl^- } (CHIEF ANIONS OF EXTRA-CELLULAR FLUID)

$HPO_4^=$
$SO_4^=$
Protein⁻ } CHIEF ANIONS OF INTRACELLULAR FLUID.

HCO_3^- ← → HCO_3^- ← HCO_3^- }

RED CELLS 45% of Total Blood Volume

H_2O
Cl^-
HCO_3^-
K^+
Hb^-

Interstitial Fluid Volume includes also small amounts of fluid in LYMPHATIC SYSTEM, CEREBROSPINAL FLUID, PERICARDIAL PLEURAL and PERITONEAL } SACS

[Areas in diagram are not proportional to actual volumes]

WATER BALANCE

In health the *total amount of body water (and salt)* is kept *reasonably constant* in spite of wide fluctuations in daily intake.

A BALANCE is struck between

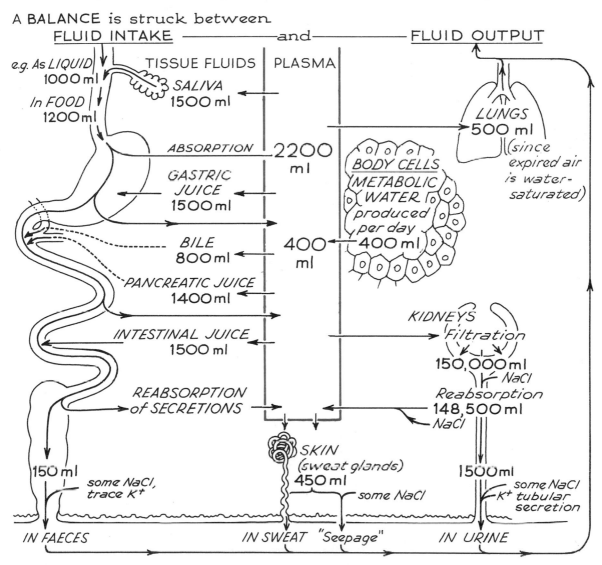

FLUID INTAKE ——————— and ——————— FLUID OUTPUT

e.g. As LIQUID 1000 ml

In FOOD 1200 ml

TISSUE FLUIDS   PLASMA

SALIVA 1500 ml

ABSORPTION 2200 ml

GASTRIC JUICE 1500 ml

BILE 800 ml

PANCREATIC JUICE 1400 ml

INTESTINAL JUICE 1500 ml

REABSORPTION of SECRETIONS

150 ml   some NaCl, trace K^+

IN FAECES

BODY CELLS METABOLIC WATER produced per day 400 ml

400 ml

LUNGS 500 ml (since expired air is water-saturated)

KIDNEYS 'Filtration 150,000 ml NaCl Reabsorption 148,500 ml NaCl

SKIN (sweat glands) 450 ml   some NaCl

1500 ml   some NaCl K^+ tubular secretion

IN SWEAT   "Seepage"   IN URINE

Except in Growth, Convalescence or Pregnancy, when new tissue is being formed, an INCREASE or DECREASE in INTAKE leads to an appropriate INCREASE or DECREASE in OUTPUT to maintain the BALANCE.

BLOOD

Blood is the specialized fluid tissue of the *TRANSPORT SYSTEM.*
[Specific Gravity, 1·055 – 1·065 ; pH, 7·3-7·4 ; Average amount, 5 litres, varies with body weight (about 7·7% of body weight).

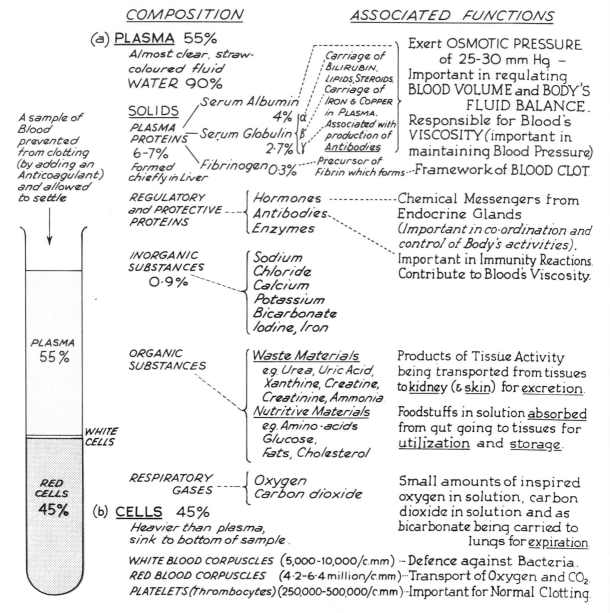

COMPOSITION

(a) PLASMA 55%
Almost clear, straw-coloured fluid
WATER 90%

SOLIDS
PLASMA PROTEINS 6-7% — Serum Albumin 4% |α
Serum Globulin 2·7% |β |γ
Formed chiefly in Liver — Fibrinogen 0·3%

REGULATORY and PROTECTIVE PROTEINS {
Hormones
Antibodies
Enzymes

INORGANIC SUBSTANCES 0·9% {
Sodium
Chloride
Calcium
Potassium
Bicarbonate
Iodine, Iron

ORGANIC SUBSTANCES {
Waste Materials
e.g. Urea, Uric Acid,
Xanthine, Creatine,
Creatinine, Ammonia
Nutritive Materials
e.g. Amino-acids
Glucose,
Fats, Cholesterol

RESPIRATORY GASES {
Oxygen
Carbon dioxide

(b) CELLS 45%
Heavier than plasma,
sink to bottom of sample.

A sample of Blood prevented from clotting (by adding an Anticoagulant) and allowed to settle

PLASMA 55%

WHITE CELLS

RED CELLS 45%

ASSOCIATED FUNCTIONS

Carriage of BILIRUBIN, LIPIDS, STEROIDS.
Carriage of IRON & COPPER in PLASMA.
Associated with production of Antibodies
Precursor of Fibrin which forms

Exert OSMOTIC PRESSURE of 25-30 mm Hg –
Important in regulating BLOOD VOLUME and BODY'S FLUID BALANCE.
Responsible for Blood's VISCOSITY (important in maintaining Blood Pressure)
Framework of BLOOD CLOT.

Chemical Messengers from Endocrine Glands
(Important in co-ordination and control of Body's activities).
Important in Immunity Reactions.
Contribute to Blood's Viscosity.

Products of Tissue Activity being transported from tissues to kidney (& skin) for excretion.

Foodstuffs in solution absorbed from gut going to tissues for utilization and storage.

Small amounts of inspired oxygen in solution, carbon dioxide in solution and as bicarbonate being carried to lungs for expiration.

WHITE BLOOD CORPUSCLES (5,000-10,000/c.mm) - Defence against Bacteria.
RED BLOOD CORPUSCLES (4·2-6·4 million/c.mm) - Transport of Oxygen and CO_2.
PLATELETS (Thrombocytes) (250,000-500,000/c.mm) - Important for Normal Clotting.

104

HAEMOSTASIS AND BLOOD COAGULATION

3 main processes are involved in HAEMOSTASIS:-
On INJURY—spasm of smooth muscle→① BLOOD VESSEL WALL CONSTRICTS.

PLATELETS stick to damaged lining ——————→ ② PLATELETS PLUG GAP

SEROTONIN – make surface of adhered platelets sticky ③ BLOOD CLOTS
and ADP   Other platelets adhere to them

BLOOD COAGULATION is initiated when
PLASMA CONTACTS DAMAGED LINING
OF BLOOD VESSEL

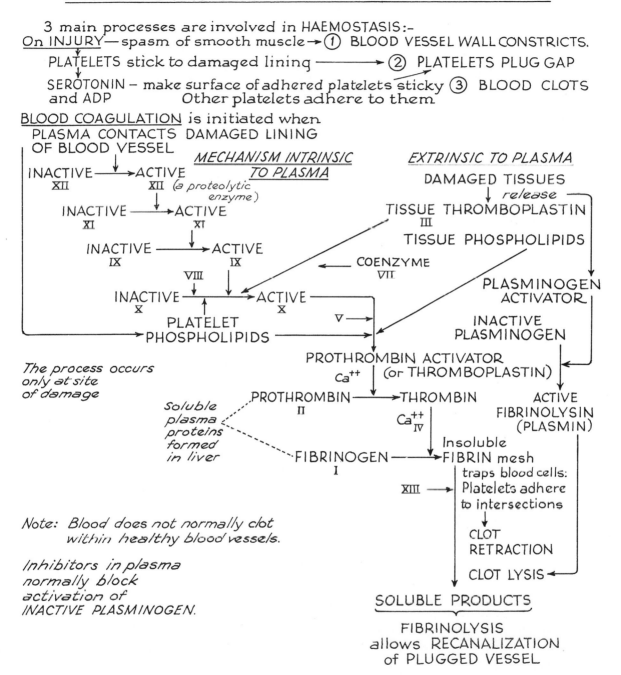

The process occurs
only at site
of damage

Note: Blood does not normally clot
   within healthy blood vessels.

Inhibitors in plasma
normally block
activation of
INACTIVE PLASMINOGEN.

FACTORS REQUIRED for NORMAL HAEMOPOIESIS

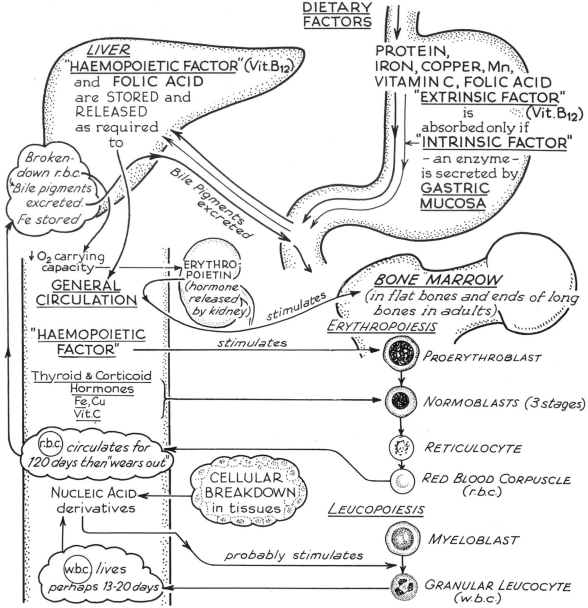

DIETARY FACTORS

LIVER
"HAEMOPOIETIC FACTOR" (Vit. B₁₂)
and FOLIC ACID
are STORED and
RELEASED
as required
to

PROTEIN,
IRON, COPPER, Mn,
VITAMIN C, FOLIC ACID
"EXTRINSIC FACTOR"
is (Vit. B₁₂)
absorbed only if
"INTRINSIC FACTOR"
– an enzyme –
is secreted by
GASTRIC MUCOSA

Broken-down r.b.c. Bile pigments excreted. Fe stored

Bile Pigments excreted

↓ O₂ carrying capacity

ERYTHRO-POIETIN (hormone released by kidney)

GENERAL CIRCULATION

stimulates

BONE MARROW
(in flat bones and ends of long bones in adults)

ERYTHROPOIESIS

"HAEMOPOIETIC FACTOR"

stimulates

PROERYTHROBLAST

Thyroid & Corticoid Hormones
Fe, Cu
Vit. C

NORMOBLASTS (3 stages)

RETICULOCYTE

r.b.c. circulates for 120 days then "wears out"

CELLULAR BREAKDOWN in tissues

RED BLOOD CORPUSCLE (r.b.c.)

NUCLEIC ACID derivatives

LEUCOPOIESIS

MYELOBLAST

w.b.c. lives perhaps 13-20 days

probably stimulates

GRANULAR LEUCOCYTE (w.b.c.)

In health the number of r.b.c. and the amount of Hb in them remain fairly constant. Destruction of old red cells is balanced by formation of new.

HAEMOPOIESIS

In the adult the formed elements of the Blood Stream develop from primitive RETICULAR CELLS chiefly in RED BONE MARROW.

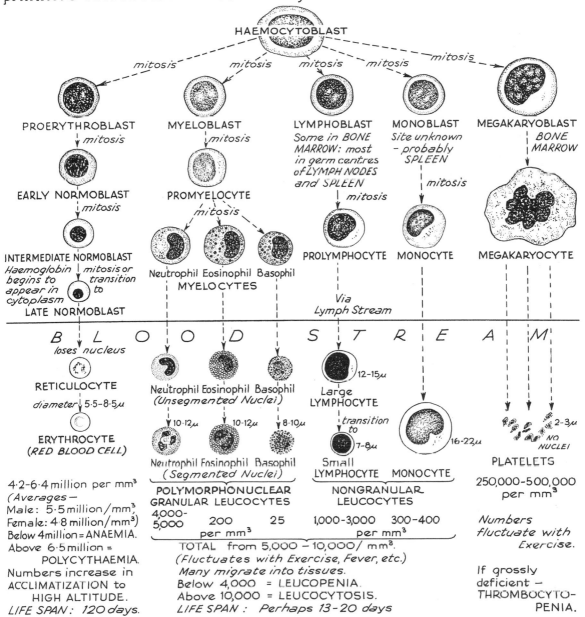

HAEMOCYTOBLAST

mitosis   *mitosis*   *mitosis*   *mitosis*   *mitosis*

PROERYTHROBLAST   MYELOBLAST   LYMPHOBLAST   MONOBLAST   MEGAKARYOBLAST

mitosis   *mitosis*   *Some in BONE MARROW: most in germ centres of LYMPH NODES and SPLEEN*   *Site unknown – probably SPLEEN*   BONE MARROW

EARLY NORMOBLAST   PROMYELOCYTE

mitosis   *mitosis*   *mitosis*   *mitosis*

INTERMEDIATE NORMOBLAST   Neutrophil Eosinophil Basophil   PROLYMPHOCYTE   MONOCYTE   MEGAKARYOCYTE

Haemoglobin begins to appear in cytoplasm   *mitosis or transition to*   MYELOCYTES

LATE NORMOBLAST

Via Lymph Stream

B L O O D S T R E A M

loses nucleus

RETICULOCYTE

Neutrophil Eosinophil Basophil *(Unsegmented Nuclei)*   Large LYMPHOCYTE 12-15μ

diameter 5.5-8.5μ   10-12μ   10-12μ   8-10μ   *transition to* 7-8μ

ERYTHROCYTE *(RED BLOOD CELL)*   Neutrophil Eosinophil Basophil *(Segmented Nuclei)*   Small LYMPHOCYTE   MONOCYTE 16-22μ   PLATELETS   2-3μ NO NUCLEI

POLYMORPHONUCLEAR GRANULAR LEUCOCYTES   NONGRANULAR LEUCOCYTES

4.2-6.4 million per mm³ *(Averages —* Male: 5.5 million/mm³, Female: 4.8 million/mm³) Below 4 million = ANAEMIA. Above 6.5 million = POLYCYTHAEMIA. Numbers increase in ACCLIMATIZATION to HIGH ALTITUDE. *LIFE SPAN: 120 days.*

4,000-5,000   200   25   1,000-3,000   300-400   250,000-500,000 per mm³

per mm³   per mm³

TOTAL from 5,000 – 10,000/ mm³. *(Fluctuates with Exercise, Fever, etc.) Many migrate into tissues.* Below 4,000 = LEUCOPENIA. Above 10,000 = LEUCOCYTOSIS. *LIFE SPAN : Perhaps 13-20 days*

Numbers fluctuate with Exercise.

If grossly deficient – THROMBOCYTO-PENIA.

BLOOD GROUPS

There are present in the PLASMA of some individuals substances which can cause the AGGLUTINATION *(clumping together)* and subsequent HAEMOLYSIS *(breakdown)* of the RED BLOOD CELLS of some other individuals.

If such reactions follow BLOOD TRANSFUSION the two bloods are said to be INCOMPATIBLE.

Two FACTORS are involved in an AGGLUTINATION reaction:-
An AGGLUTINOGEN present in DONOR'S Red Blood Cell } e.g. A } or B }
A specific AGGLUTININ present in RECIPIENT'S Plasma } α } or β }

Obviously no such combination occurs naturally otherwise auto-agglutination would result.

In the ABO System

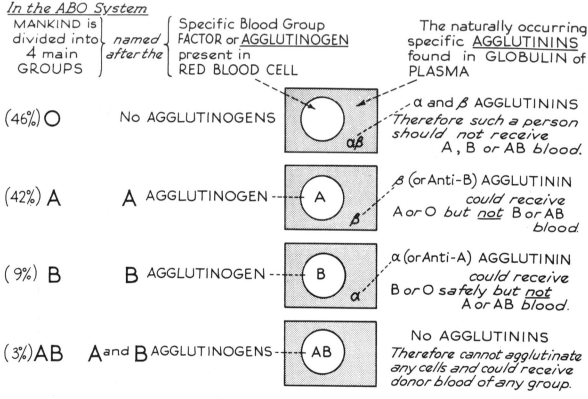

MANKIND is divided into 4 main GROUPS	*named after the*	Specific Blood Group FACTOR or AGGLUTINOGEN present in RED BLOOD CELL	The naturally occurring specific AGGLUTININS found in GLOBULIN of PLASMA
(46%) O		No AGGLUTINOGENS	α and β AGGLUTININS *Therefore such a person should not receive A, B or AB blood.*
(42%) A		A AGGLUTINOGEN	β (or Anti-B) AGGLUTININ *could receive A or O but not B or AB blood.*
(9%) B		B AGGLUTINOGEN	α (or Anti-A) AGGLUTININ *could receive B or O safely but not A or AB blood.*
(3%) AB		A and B AGGLUTINOGENS	No AGGLUTININS *Therefore cannot agglutinate any cells and could receive donor blood of any group.*

In Practice *it is important that the* DONOR's *cells should not be agglutinated by the* RECIPIENT's *plasma. Agglutination of Recipient's cells by Donor agglutinins is less likely to occur.*

There are sub-groups within these main groups. These have to be considered in blood transfusion [work.

BLOOD GROUPS

To determine the Blood Group to which an individual belongs *TWO TEST SERA* only are required and the *RED BLOOD CELLS* to be grouped.

DROP of <u>GROUP A SERUM</u> & <u>GROUP B SERUM</u>
(on glass slide)   [N.B. These droplets are of PLASMA – <u>NOT</u> Red Blood Cells. i.e. only AGGLUTININS are present.]

To each test serum a saline suspension of RED BLOOD CELLS is added. *i.e. only AGGLUTINOGENS are added.*

<u>GROUP A SERUM</u> | <u>GROUP B SERUM</u>

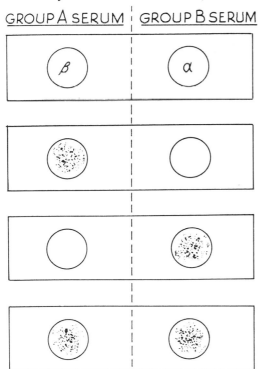

GROUP **O** Blood Cells give <u>NO</u> agglutination since <u>NO AGGLUTINOGENS</u> are present in these cells to be clumped by test sera AGGLUTININS.

GROUP **B** Blood Cells (B AGGLUTINOGEN present) give agglutination with GROUP **A** serum – since the specific Anti-B (β) AGGLUTININ is present in the first test serum.

GROUP **A** Blood Cells give agglutination with GROUP **B** serum — since the specific Anti-A (α) AGGLUTININ is present in this serum.

GROUP **AB** Blood Cells give agglutination with both test sera.

As well as determining Blood Group in this way, the Blood of Donor is always matched directly with Blood of Patient to avoid sub-group incompatibility.

AGGLUTINATION is usually visible under the microscope within a few minutes. The clumped cells look like grains of cayenne pepper in a clear liquid. If no agglutination occurs the fluid remains uniformly pink.

If the wrong blood is given to a patient, clumps of red blood cells may block small blood vessels in vital organs, e.g. lung or brain. The subsequent haemolysis (breakdown) of agglutinated cells may lead to Haemoglobin in the Urine and eventually to Kidney Failure and Death.

RHESUS FACTOR

Over 50 different BLOOD GROUP FACTORS have been demonstrated. Not all are important in blood transfusion work.

The <u>RHESUS FACTOR</u> is important

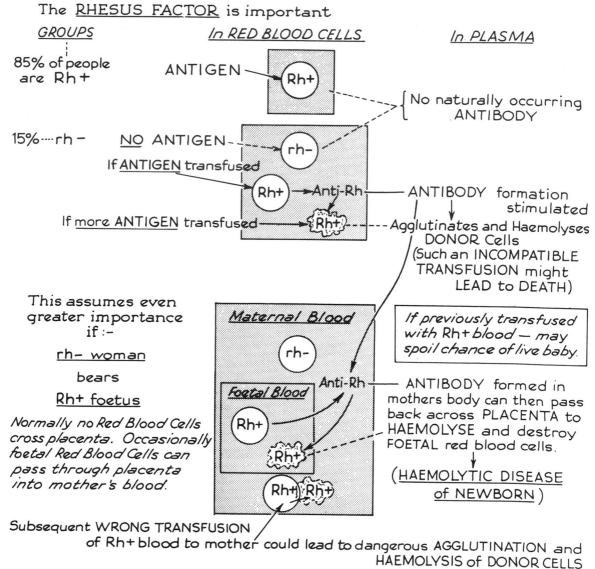

GROUPS

85% of people are Rh+

15%····rh−

In RED BLOOD CELLS

ANTIGEN

<u>NO</u> ANTIGEN

If <u>ANTIGEN</u> transfused

Anti-Rh

If <u>more ANTIGEN</u> transfused

In PLASMA

No naturally occurring ANTIBODY

ANTIBODY formation stimulated

Agglutinates and Haemolyses DONOR Cells
(Such an INCOMPATIBLE TRANSFUSION might LEAD to DEATH)

This assumes even greater importance if :-

rh− woman

bears

Rh+ foetus

Normally no Red Blood Cells cross placenta. Occasionally foetal Red Blood Cells can pass through placenta into mother's blood.

Maternal Blood

Foetal Blood

Anti-Rh

If previously transfused with Rh+ blood — may spoil chance of live baby.

ANTIBODY formed in mothers body can then pass back across PLACENTA to HAEMOLYSE and destroy FOETAL red blood cells.

(<u>HAEMOLYTIC DISEASE of NEWBORN</u>)

Subsequent WRONG TRANSFUSION
of Rh+ blood to mother could lead to dangerous AGGLUTINATION and HAEMOLYSIS of DONOR CELLS within mother's own body.

There are many sub-groups in this system.

INHERITANCE of RHESUS FACTOR

This BLOOD GROUP FACTOR is inherited on Mendelian principles.

HOMOZYGOUS
Rh Positive
FATHER

LACK of the FACTOR is a
Mendelian recessive
rh negative
MOTHER

All Spermatozoa carry the FACTOR: *All Ova lack the FACTOR*

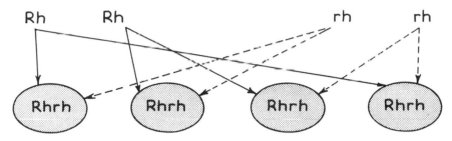

All children will be Rhrh, *i.e.* all carry the FACTOR and are Rh Positive. *i.e.* During pregnancy the rh negative woman will have a Rh Positive foetus.

HETEROZYGOUS
Rh Positive
FATHER

rh negative
MOTHER

Half the Spermatozoa carry the FACTOR: *All Ova lack the FACTOR*

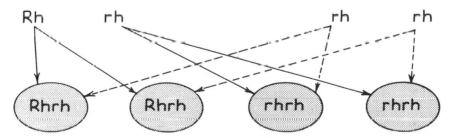

Possibility that some children will be Rhrh Positive, like the father; others rhrh negative like the mother.

LYMPHATIC SYSTEM

All CELLS
are bathed by TISSUE FLUID.
This diffuses from CAPILLARIES.
Some returns to CAPILLARIES.
Some drains into blind-ending,
thin-walled LYMPHATICS.
It is then known as LYMPH
(similar to plasma but less protein).

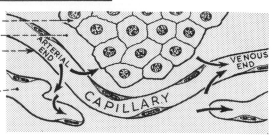

A network of LYMPHATIC VESSELS drains tissue spaces throughout the body *(except in Central Nervous System).* They unite to form LARGER and LARGER vessels ⟶ RIGHT LYMPHATIC DUCT and LEFT THORACIC DUCT ⟶ SUBCLAVIAN VEINS *(i.e. lymph is returned to the blood stream here)*

In the course of LARGER vessels, LYMPH is filtered through <u>LYMPH NODES</u>

AFFERENT LYMPHATICS – pour their LYMPH into RETICULAR FRAMEWORK of loose Sinus tissue.

MACROPHAGE cells ingest foreign material (e.g. carbon in lungs) or harmful bacteria.

LYMPH NODULES produce LYMPHOCYTES and PLASMA CELLS

important in antibody formation and immuno-logical reactions

Capsule of Fibrous Tissue

EFFERENT LYMPHATIC — receives lymph after its slow passage through node.

MOVEMENT of LYMPH towards HEART depends partly on compression of lymphatic vessels by muscles of limbs and partly on 'suction' created by movements of respiration. Valves within the vessels prevent backflow. The lymphatic tissue of the body forms an important part of the Body's Defence against invading agents such as PROTOZOA, BACTERIA, VIRUSES, or their poisonous TOXINS. These act as ANTIGENS stimulating ANTIBODY FORMATION —— which can subsequently destroy or neutralize the antigen.

SPLEEN

The Spleen is a vascular organ, weighing about 200 grams. It is situated in the left side of the abdomen behind the stomach and above the kidney.

FIBROUS TISSUE CAPSULE
with extensions into substance of organ
TRABECULAE

SPLENIC ARTERY and VEIN
and branches
TRABECULAR ARTERY and VEIN

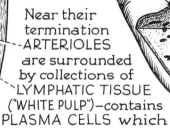

Near their termination ARTERIOLES are surrounded by collections of LYMPHATIC TISSUE ("WHITE PULP")–contains PLASMA CELLS which manufacture many antibodies and produces LYMPHOCYTES → added to blood in its passage through spleen.

"RED PULP"– Framework of Reticular Tissue — acts as RESERVOIR for Blood.

(Some smooth muscle in Capsule and Trabeculae may aid in expelling blood from reservoir to general circulation.)

| PHAGOCYTIC CELLS of ELLIPSOID and other Reticulo-Endothelial Cells | Destroy worn out Red Blood Corpuscles, platelets and other (foreign) particles, etc. |

It is difficult to trace further pathways of blood through spleen and there is much disagreement about the exact route taken. Probably Arterioles link up with VENOUS SINUSES (in the way shown).

Blood appears to percolate through the narrow spaces in the wall of the SINUS into the Red Pulp — and can get back into Sinus in same way.

During foetal life the spleen forms red and white blood cells.
It is not apparently essential to life in the adult.

113

CEREBROSPINAL FLUID

COMPOSITION:- Cerebrospinal Fluid (C.S.F.) is similar to blood plasma but does not clot. It is a clear watery alkaline fluid. It contains salts, glucose, some urea and creatinine, very little protein and very few lymphocytes.
VOLUME:- 120-150 ml. in man.   SPECIFIC GRAVITY:- 1·005-1·008.

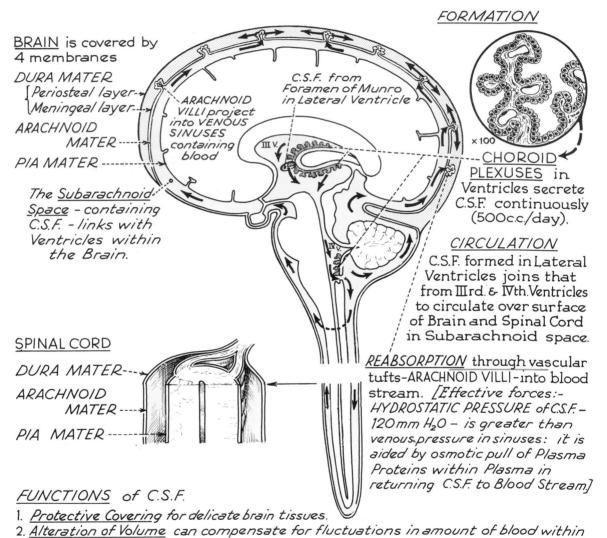

FORMATION

BRAIN is covered by 4 membranes

DURA MATER
{ Periosteal layer-->
{ Meningeal layer-->

ARACHNOID MATER

PIA MATER

ARACHNOID VILLI project into VENOUS SINUSES containing blood

C.S.F. from Foramen of Munro in Lateral Ventricle

III V.

IV V.

The Subarachnoid Space - containing C.S.F. - links with Ventricles within the Brain.

× 100

CHOROID PLEXUSES in Ventricles secrete C.S.F. continuously (500 c.c./day).

CIRCULATION
C.S.F. formed in Lateral Ventricles joins that from IIIrd. & IVth.Ventricles to circulate over surface of Brain and Spinal Cord in Subarachnoid space.

SPINAL CORD

DURA MATER---->

ARACHNOID MATER---->

PIA MATER------>

REABSORPTION through vascular tufts-ARACHNOID VILLI-into blood stream. [Effective forces:- HYDROSTATIC PRESSURE of C.S.F.- 120 mm H_2O – is greater than venous pressure in sinuses: it is aided by osmotic pull of Plasma Proteins within Plasma in returning C.S.F. to Blood Stream]

FUNCTIONS of C.S.F.
1. Protective Covering for delicate brain tissues.
2. Alteration of Volume can compensate for fluctuations in amount of blood within skull and thus keep total volume of cranial contents constant.
3. Exchange of Metabolic substances between Nerve Cells and C.S.F.
(i.e. it receives some waste products.)

114

CHAPTER 5

RESPIRATORY SYSTEM

RESPIRATORY SYSTEM

All living cells require to get OXYGEN from the fluid around them and to get rid of CARBON DIOXIDE to it.

<u>INTERNAL RESPIRATION</u> is the exchange of these gases between tissue cells and their fluid environment.

<u>EXTERNAL RESPIRATION</u> is the exchange of these gases *(oxygen and carbon dioxide)* between the body and the external environment.

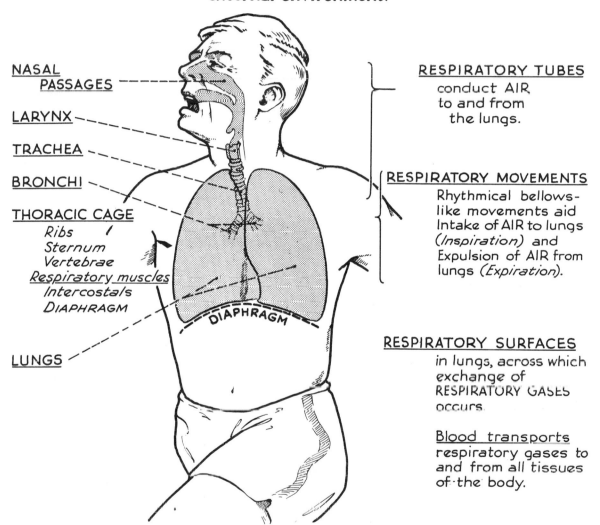

<u>NASAL PASSAGES</u>

<u>LARYNX</u>

<u>TRACHEA</u>

<u>BRONCHI</u>

<u>THORACIC CAGE</u>
Ribs
Sternum
Vertebrae
<u>*Respiratory muscles*</u>
Intercostals
DIAPHRAGM

<u>LUNGS</u>

DIAPHRAGM

<u>RESPIRATORY TUBES</u>
conduct AIR
to and from
the lungs.

<u>RESPIRATORY MOVEMENTS</u>
Rhythmical bellows-like movements aid Intake of AIR to lungs *(Inspiration)* and Expulsion of AIR from lungs *(Expiration)*.

<u>RESPIRATORY SURFACES</u>
in lungs, across which exchange of RESPIRATORY GASES occurs.

<u>Blood transports</u> respiratory gases to and from all tissues of the body.

117

AIR CONDUCTING PASSAGES

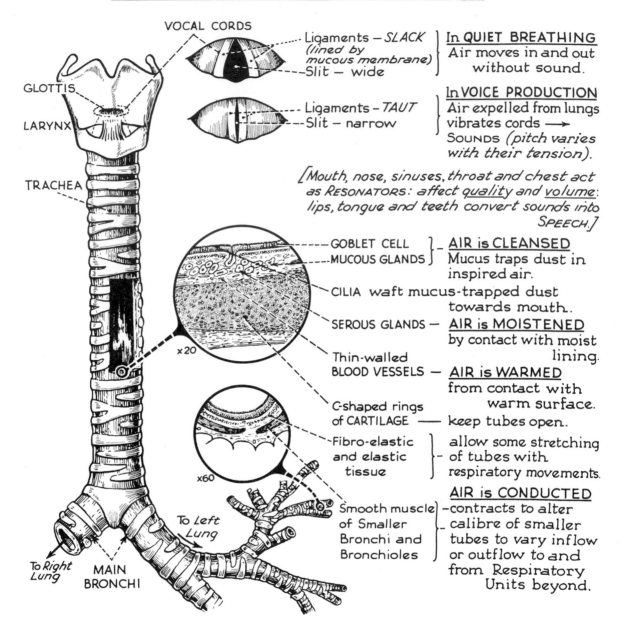

VOCAL CORDS

GLOTTIS

LARYNX

TRACHEA

-- Ligaments — *SLACK*
(lined by
mucous membrane)
-- Slit — wide

In QUIET BREATHING
Air moves in and out
without sound.

-- Ligaments – *TAUT*
-- Slit – narrow

In VOICE PRODUCTION
Air expelled from lungs
vibrates cords ⟶
SOUNDS *(pitch varies*
with their tension).

[Mouth, nose, sinuses, throat and chest act
as RESONATORS: affect quality and volume:
lips, tongue and teeth convert sounds into
SPEECH.]

-- GOBLET CELL
-- MUCOUS GLANDS

AIR is CLEANSED
Mucus traps dust in
inspired air.

-- CILIA waft mucus-trapped dust
towards mouth.

-- SEROUS GLANDS — **AIR is MOISTENED**
by contact with moist
lining.

-- Thin-walled
BLOOD VESSELS — **AIR is WARMED**
from contact with
warm surface.

x20

-- C-shaped rings
of CARTILAGE —— keep tubes open.

-- Fibro-elastic
and elastic
tissue

allow some stretching
of tubes with
respiratory movements.

x60

AIR is CONDUCTED

-- Smooth muscle
of Smaller
Bronchi and
Bronchioles

-contracts to alter
calibre of smaller
tubes to vary inflow
or outflow to and
from Respiratory
Units beyond.

To *Left*
Lung

To *Right*
Lung
MAIN
BRONCHI

LUNGS: RESPIRATORY SURFACES

The Trachea and the Bronchial 'Tree' conduct Air down to the
RESPIRATORY SURFACES.

*There is no exchange of gases in
these tubes.*

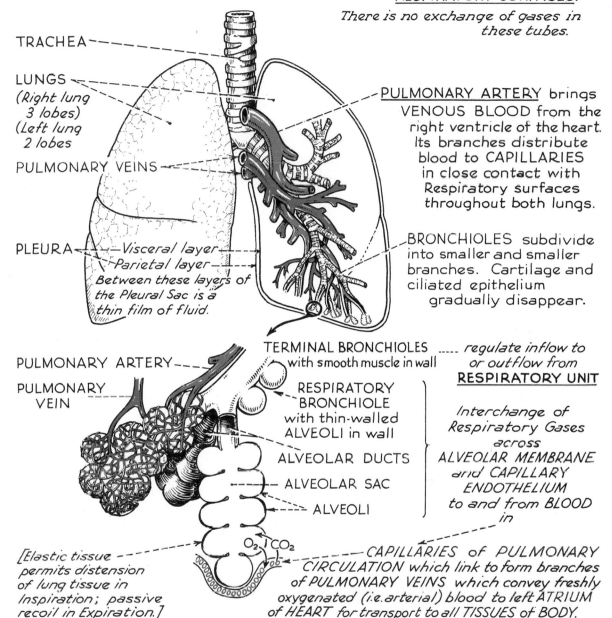

TRACHEA

LUNGS
*(Right lung
3 lobes)*
*(Left lung
2 lobes*

PULMONARY VEINS

PLEURA —— *Visceral layer*
Parietal layer
*Between these layers of
the Pleural Sac is a
thin film of fluid.*

PULMONARY ARTERY brings
VENOUS BLOOD from the
right ventricle of the heart.
Its branches distribute
blood to CAPILLARIES
in close contact with
Respiratory surfaces
throughout both lungs.

BRONCHIOLES subdivide
into smaller and smaller
branches. Cartilage and
ciliated epithelium
gradually disappear.

TERMINAL BRONCHIOLES ..... *regulate inflow to
with smooth muscle in wall* *or outflow from*
RESPIRATORY UNIT

PULMONARY ARTERY

PULMONARY
VEIN

RESPIRATORY
BRONCHIOLE
with thin-walled
ALVEOLI in wall

ALVEOLAR DUCTS

ALVEOLAR SAC

ALVEOLI

*Interchange of
Respiratory Gases
across
ALVEOLAR MEMBRANE
and CAPILLARY
ENDOTHELIUM
to and from BLOOD
in*

O_2 CO_2

*[Elastic tissue
permits distension
of lung tissue in
Inspiration; passive
recoil in Expiration.]*

*CAPILLARIES of PULMONARY
CIRCULATION which link to form branches
of PULMONARY VEINS which convey freshly
oxygenated (i.e. arterial) blood to left ATRIUM
of HEART for transport to all TISSUES of BODY.*

119

THORAX

The Thorax (*or* chest) is the closed cavity which contains the LUNGS, HEART and Great Vessels.

It is enclosed and bounded
<u>ABOVE</u> by the upper RIBS and tissues of the neck;
<u>AT THE SIDES</u> by the RIBS and INTERCOSTAL MUSCLES;
<u>AT THE BACK</u> by the RIBS and VERTEBRAL COLUMN (or *back bone*);
<u>IN FRONT</u> by the RIBS, COSTAL CARTILAGES and STERNUM (*or breast bone*);
<u>BELOW</u> by the DIAPHRAGM (*a strong dome-shaped sheet of skeletal muscle which separates the thoracic cavity from the abdominal cavity*).

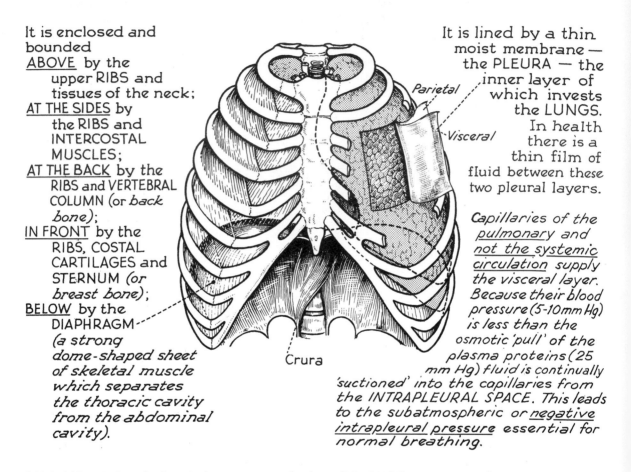

Parietal

Visceral

Crura

It is lined by a thin moist membrane — the PLEURA — the inner layer of which invests the LUNGS. In health there is a thin film of fluid between these two pleural layers.

Capillaries of the <u>pulmonary</u> and <u>not the systemic circulation</u> supply the visceral layer. Because their blood pressure (5-10 mm Hg) is less than the osmotic 'pull' of the plasma proteins (25 mm Hg) fluid is continually 'suctioned' into the capillaries from the INTRAPLEURAL SPACE. This leads to the subatmospheric or <u>negative intrapleural pressure</u> essential for normal breathing.

DIMENSIONS of thoracic cage and the PRESSURE between pleural surfaces change rhythmically about 18-20 times a minute with the MOVEMENTS of RESPIRATION —— AIR MOVEMENT in and out of the lungs follows
passively.

MECHANISM of BREATHING

The rhythmical changes in the CAPACITY of the Thorax are brought about by MUSCULAR ACTION. The changes in LUNG VOLUME with INTAKE or EXPULSION of air follow passively.

In NORMAL QUIET BREATHING

INSPIRATION

EXTERNAL INTERCOSTAL MUSCLES actively contract-
- ribs and sternum move upwards and outwards
- width of chest increases from side to side and from front to back.

DIAPHRAGM contracts -
- descends
- depth of chest increases.

CAPACITY of THORAX is INCREASED

↓

PRESSURE between PLEURAL SURFACES (already negative) is REDUCED from -2 to -6 mmHg (i.e. an increased "suction pull" is exerted on LUNG TISSUE)

↓

ELASTIC TISSUE of LUNGS is STRETCHED

↓

LUNGS EXPAND to fill THORACIC CAVITY

↓

AIR PRESSURE within ALVEOLI is now less than atmospheric pressure

↓

AIR is sucked into ALVEOLI from ATMOSPHERE

EXPIRATION

EXTERNAL INTERCOSTAL MUSCLES relax —
- ribs and sternum move downwards and inwards
- width of chest diminishes.

DIAPHRAGM relaxes —
- ascends
- depth of chest diminishes.

CAPACITY of THORAX is DECREASED

↓

PRESSURE between PLEURAL SURFACES is INCREASED from -6 to -2 mm Hg (i.e. less pull is exerted on LUNG TISSUE)

↓

ELASTIC TISSUE of LUNGS RECOILS

↓

AIR PRESSURE within ALVEOLI is now greater than atmospheric pressure

↓

AIR is forced out of ALVEOLI to ATMOSPHERE

In FORCED BREATHING

Muscles of Nostrils and round Glottis may contract to aid entrance of air to lungs.
Extensors of Vertebral Column may aid inspiration.
Muscles of Neck contract-
- move 1ST rib upwards
(and sternum upwards and forwards)

Internal Intercostals may contract-
- move ribs downwards more actively.
Abdominal muscles contract-
- actively aid ascent of diaphragm.

121

ARTIFICIAL RESPIRATION

Many methods aim at alternately DECREASING and INCREASING the CAPACITY of the THORAX 15-18 times per minute in imitation of the normal expiratory and inspiratory movements.

Two well-known methods illustrate the mechanism:-
[In each case mouth and throat must be clear of obstruction and tongue drawn forwards]

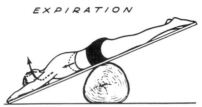
EXPIRATION

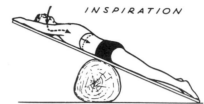
INSPIRATION

1. EVE'S ROCKING STRETCHER — uses the abdominal organs like a piston against the diaphragm.

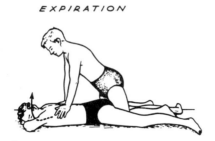
EXPIRATION

INSPIRATION

2. SCHÄFER —— pressure applied to compress lower part of rib cage and so decrease capacity of thorax — *AIR MOVES OUT.*
On release of pressure thorax returns to previous position and capacity — *AIR SUCKED IN.*

In the MOUTH-to-MOUTH method the applicator pinches shut the patient's nostrils, seals his lips round the patient's mouth and blows in air [C]. When he removes his mouth the patient breathes out passively [D]. *[The residual oxygen in the applicator's own expired air is thus made available to patient]*

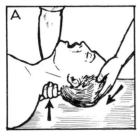 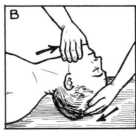 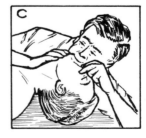 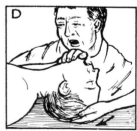

CAPACITY of LUNGS

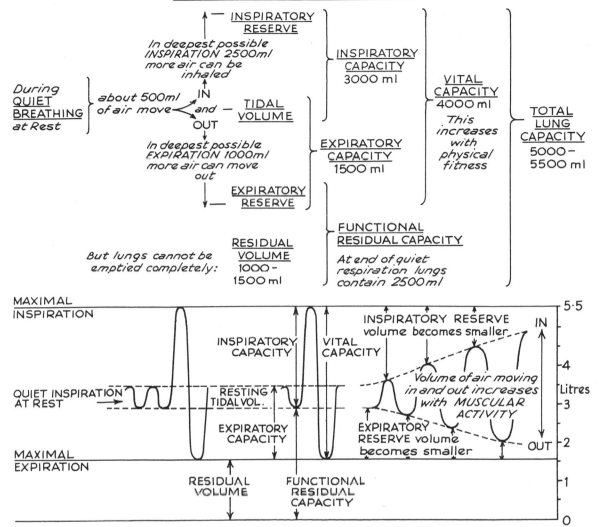

(After PAPPENHEIMER, J.R., et al (1950) Fed. Proc., 9,602)

At rest a normal male adult breathes in and out about 16 times per minute. The amount of air breathed in per minute is therefore 500ml × 16, i.e. 8000ml or 8 litres — This is the RESPIRATORY MINUTE VOLUME or PULMONARY VENTILATION.   In exercise it may go up to as much as 200 litres.   These values are about 25% lower in women.

In deep breathing the volume of ATMOSPHERIC AIR INSPIRED with each inspiration and the amount which reaches the ALVEOLI increase.

COMPOSITION of RESPIRED AIR

In *QUIET BREATHING*

Of the
500 ml ATMOSPHERIC AIR
INSPIRED in a single inspiration
    OXYGEN *makes up* 20·95%
    NITROGEN ——— 79·01%
    CARBON DIOXIDE— 0·04%

Of the AIR
EXPIRED in a single expiration
    OXYGEN *makes up* 16·4%
    NITROGEN ——— 79·6%
    CARBON DIOXIDE — 4·0%

[N.B. Nitrogen is <u>not</u> given out by body.
The Volume of CO_2 given up by the
blood is slightly less than the
Volume of O_2 taken up by the blood.
The Volume of air expired is therefore
less than Volume inspired.
Both contain the same <u>absolute</u>
amount of nitrogen.]

140 ml occupy the CONDUCTING
PASSAGES —"DEAD SPACE"
AIR.　This remains
unchanged in composition
since it is not in contact
with Respiratory Surfaces.

This represents a mixture of-
-"DEAD SPACE" AIR – air which
has moved out *unchanged*
from the CONDUCTING
    PASSAGES

and

360ml *(together with 140ml*
previously occupying
conducting passages)
reach the ALVEOLI to mix
with air already in
contact with
Respiratory Surfaces.
This mixture changes
composition as
it gives up
OXYGEN to the blood
and takes up
CARBON DIOXIDE
from the blood

ALVEOLAR AIR – air which
has been in contact with
Respiratory Surfaces and
has given up some OXYGEN
to the blood and taken up
CARBON DIOXIDE from it.
    OXYGEN ——— 13·8%
    NITROGEN ——— 80·7%
    CARBON DIOXIDE – 5·5%

With each *EXPIRATION* 140ml
ALVEOLAR AIR move up to fill
Conducting passages.
This is drawn back into
ALVEOLI with start of next
    INSPIRATION.

O_2
CO_2
BLOOD

In <u>DEEP BREATHING</u>　the composition of Air breathed out with each
    expiration comes closer to that of Alveolar Air.

MOVEMENT of RESPIRATORY GASES

A gas moves from an area where it is present at *higher concentration* to an area where it is present at *lower concentration*. The movement of gas molecules continues till the pressure exerted by them is the same throughout both areas. DRY Atmospheric Air *(at sea level)* has a pressure of **760 mm Hg**

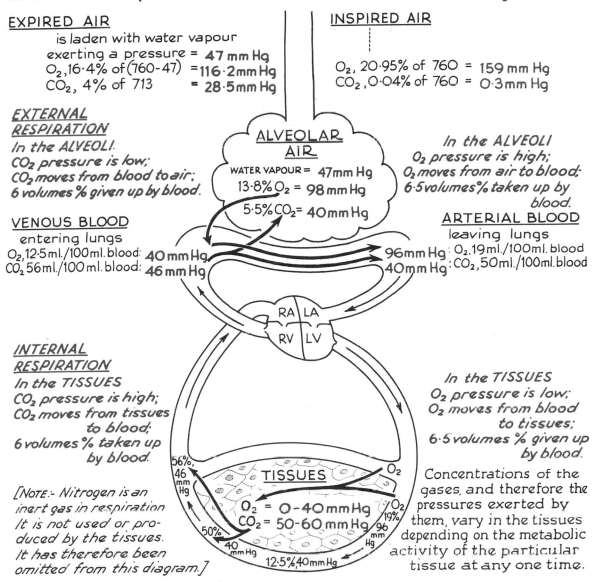

EXPIRED AIR
is laden with water vapour
exerting a pressure = **47 mm Hg**
O_2, 16·4% of (760-47) = **116·2 mm Hg**
CO_2, 4% of 713 = **28·5 mm Hg**

INSPIRED AIR
O_2, 20·95% of 760 = **159 mm Hg**
CO_2, 0·04% of 760 = **0·3 mm Hg**

EXTERNAL RESPIRATION
In the ALVEOLI.
CO_2 pressure is low;
CO_2 moves from blood to air;
6 volumes % given up by blood.

ALVEOLAR AIR
WATER VAPOUR = **47 mm Hg**
13·8% O_2 = **98 mm Hg**
5·5% CO_2 = **40 mm Hg**

In the ALVEOLI
O_2 pressure is high;
O_2 moves from air to blood;
6·5 volumes% taken up by blood.

VENOUS BLOOD
entering lungs
O_2, 12·5 ml./100 ml. blood: **40 mm Hg**
CO_2, 56 ml./100 ml. blood: **46 mm Hg**

ARTERIAL BLOOD
leaving lungs
96 mm Hg: O_2, 19 ml./100 ml. blood
40 mm Hg: CO_2, 50 ml./100 ml. blood

RA | LA
RV | LV

INTERNAL RESPIRATION
In the TISSUES
CO_2 pressure is high;
CO_2 moves from tissues to blood;
6 volumes % taken up by blood.

In the TISSUES
O_2 pressure is low;
O_2 moves from blood to tissues;
6·5 volumes % given up by blood.

[NOTE:- Nitrogen is an inert gas in respiration It is not used or produced by the tissues. It has therefore been omitted from this diagram.]

TISSUES
56%, 46 mm Hg
O_2
O_2 = 0-40 mm Hg
CO_2 = 50-60 mm Hg
O_2 19% 96 mm Hg
50%
40 mm Hg
12·5%, 40 mm Hg

Concentrations of the gases, and therefore the pressures exerted by them, vary in the tissues depending on the metabolic activity of the particular tissue at any one time.

DISSOCIATION of OXYGEN from HAEMOGLOBIN

The amount of O_2 taken up by HAEMOGLOBIN in the LUNGS or given up by OXYHAEMOGLOBIN in the TISSUES depends on the PARTIAL PRESSURE of the O_2 in the immediate environment.

It is also influenced by the PARTIAL PRESSURE of CO_2, by TEMPERATURE and by ACIDITY.

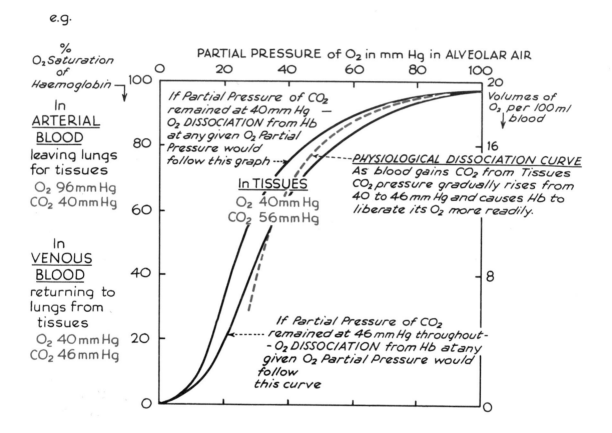

e.g.

%
O_2 Saturation
of
Haemoglobin →

In
ARTERIAL
BLOOD
leaving lungs
for tissues
O_2 96 mm Hg
CO_2 40 mm Hg

In
VENOUS
BLOOD
returning to
lungs from
tissues
O_2 40 mm Hg
CO_2 46 mm Hg

PARTIAL PRESSURE of O_2 in mm Hg in ALVEOLAR AIR

If Partial Pressure of CO_2 remained at 40 mm Hg — O_2 DISSOCIATION from Hb at any given O_2 Partial Pressure would follow this graph ···→

In TISSUES
O_2 40 mm Hg
CO_2 56 mm Hg

Volumes of
O_2 per 100 ml
↓ blood

PHYSIOLOGICAL DISSOCIATION CURVE
As blood gains CO_2 from Tissues CO_2 pressure gradually rises from 40 to 46 mm Hg and causes Hb to liberate its O_2 more readily.

If Partial Pressure of CO_2 remained at 46 mm Hg throughout - O_2 DISSOCIATION from Hb at any given O_2 Partial Pressure would follow this curve

This effect of CO_2 partial pressure on dissociation of O_2 from Hb (the Böhr effect) is advantageous.

e.g. An increase in CO_2 partial pressure locally during tissue activity causes Hb to part more readily with its O_2 to the active tissues.

UPTAKE and RELEASE of CARBON DIOXIDE

CO_2 is carried by both red blood corpuscles and plasma. The bulk of it is carried as BICARBONATE formed chiefly in red cells and carried largely by plasma. The amount of CO_2 taken up by the blood in the tissues or released by the blood in the lungs depends on the PARTIAL PRESSURES of the CO_2 in the immediate environment and also on the state of the HAEMOGLOBIN or on the amounts of OXYGEN linked to Hb at any one time.

In R.B.C.		In Plasma
Hb ——————— and ———	Plasma Proteins act as *weak acids.*
KHb *(reduced Hb)* — and ———	Na Proteinate are salts of these acids.
K^+——————— and ———	Na^+ are easily displaced by stronger acids
	(e.g. by H_2CO_3 formed from union of $CO_2 + H_2O$)

$KHbO_2$ *(oxygenated Hb)* is a *stronger* acid and
K^+ is <u>less</u> readily displaced from it by H_2CO_3.

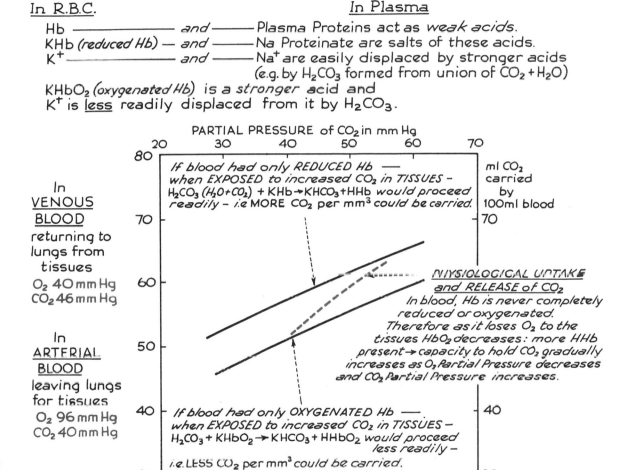

PARTIAL PRESSURE of CO_2 in mm Hg

In
<u>VENOUS
BLOOD</u>
returning to
lungs from
tissues
O_2 40 mm Hg
CO_2 46 mm Hg

In
<u>ARTERIAL
BLOOD</u>
leaving lungs
for tissues
O_2 96 mm Hg
CO_2 40 mm Hg

*If blood had only REDUCED Hb —
when EXPOSED to increased CO_2 in TISSUES –
H_2CO_3 $(H_2O + CO_2)$ + KHb → $KHCO_3$ + HHb would proceed
readily – i.e MORE CO_2 per mm^3 could be carried.*

ml CO_2
carried
by
100ml blood

<u>PHYSIOLOGICAL UPTAKE</u>
<u>and RELEASE of CO_2</u>
*In blood, Hb is never completely
reduced or oxygenated.
Therefore as it loses O_2 to the
tissues HbO_2 decreases: more HHb
present → capacity to hold CO_2 gradually
increases as O_2 Partial Pressure decreases
and CO_2 Partial Pressure increases.*

*If blood had only OXYGENATED Hb —
when EXPOSED to increased CO_2 in TISSUES –
H_2CO_3 + $KHbO_2$ → $KHCO_3$ + $HHbO_2$ would proceed
less readily –
i.e. LESS CO_2 per mm^3 could be carried.*

*i.e. The more OXYGEN the blood holds the less CO_2 it can hold and vice-versa.
This facilitates release of CO_2 in lungs and uptake of CO_2 in tissues.*

CARRIAGE and TRANSFER of OXYGEN and CARBON DIOXIDE

When ARTERIAL BLOOD is delivered by SYSTEMIC CAPILLARIES to the TISSUES it is exposed to:-

| OXYGEN at very low tensions (1-40 mm Hg) | CARBON DIOXIDE at high tensions (50-60 mm Hg) |

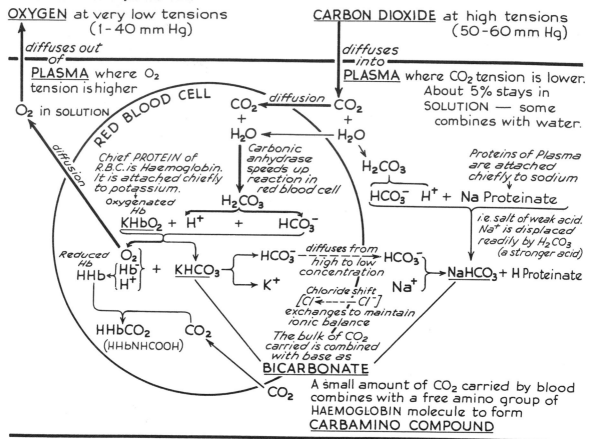

OXYGEN at very low tensions (1-40 mm Hg)

diffuses out of

PLASMA where O_2 tension is higher

O_2 in SOLUTION

RED BLOOD CELL

Chief PROTEIN of R.B.C. is Haemoglobin. It is attached chiefly to potassium.

Oxygenated Hb

$KHbO_2 + H^+ + HCO_3^-$

Reduced Hb
$HHb \{ Hb^- \\ H^+ \} + KHCO_3$

$HHbCO_2$
(HHbNHCOOH)

CO_2

CO_2

CARBON DIOXIDE at high tensions (50-60 mm Hg)

diffuses into

PLASMA where CO_2 tension is lower. About 5% stays in SOLUTION — some combines with water.

CO_2
+
H_2O
diffusion
CO_2
+
H_2O

H_2CO_3

Carbonic anhydrase speeds up reaction in red blood cell

H_2CO_3

Proteins of Plasma are attached chiefly to sodium

$HCO_3^- \quad H^+ + Na\ Proteinate$

i.e. salt of weak acid. Na^+ is displaced readily by H_2CO_3 (a stronger acid)

$HCO_3^- —$ diffuses from high to low concentration HCO_3^-

K^+

Chloride shift [$Cl^- \longleftarrow -- Cl^-$] exchanges to maintain ionic balance

Na^+

$NaHCO_3 + H\ Proteinate$

The bulk of CO_2 carried is combined with base as
BICARBONATE

A small amount of CO_2 carried by blood combines with a free amino group of HAEMOGLOBIN molecule to form **CARBAMINO COMPOUND**

During its passage through the Tissues blood gives up about 5 - 7 volumes % OXYGEN, i.e. its Hb is still up to 70% O_2 saturated.

The release of O_2 to tissues is speeded up by an increase in temperature or acidity such as occurs when tissues are active.

During its passage through the Tissues blood takes up about 4 to 6 volumes % CO_2.

Of the total CO_2 carried by blood, about 30% is carried in R.B.C., 70% in Plasma. Most of the BICARBONATE is _formed_ in R.B.C. but the HCO_3^- diffusion gradient caused by carbonic anhydrase means that most of it is _carried_ in plasma.

CARRIAGE and TRANSFER of OXYGEN and CARBON DIOXIDE

When <u>VENOUS BLOOD</u> flows through the <u>PULMONARY CAPILLARIES</u> it is exposed to:-

<u>OXYGEN</u> at high pressures
in ALVEOLAR AIR   (98 mm Hg)

<u>CARBON DIOXIDE</u> at low pressures
(40 mm Hg)

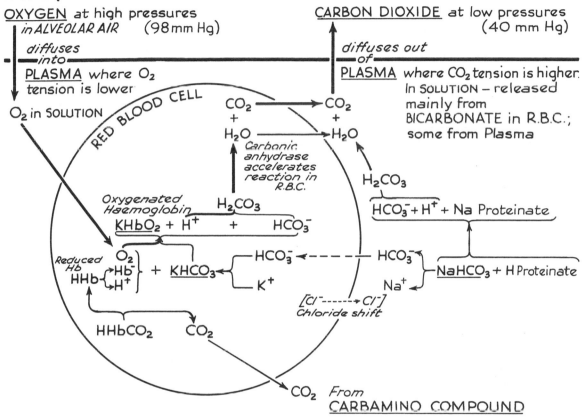

diffuses into
<u>PLASMA</u> where O_2 tension is lower

O_2 in SOLUTION

RED BLOOD CELL

diffuses out of
<u>PLASMA</u> where CO_2 tension is higher. In SOLUTION – released mainly from BICARBONATE in R.B.C.; some from Plasma

CO_2 ⟶ CO_2
+   +
H_2O ⟶ H_2O

Carbonic anhydrase accelerates reaction in R.B.C.

H_2CO_3

Oxygenated Haemoglobin
$KHbO_2 + H^+$ + HCO_3^- H_2CO_3

$HCO_3^- + H^+ + Na$ Proteinate

Reduced Hb
$HHb \{ Hb^- \\ H^+ \}$ O_2 $+ \underline{KHCO_3}$ HCO_3^- K^+ HCO_3^- Na^+ $\underline{NaHCO_3} + H$ Proteinate

$[Cl^- \cdots\!\rightarrow Cl^-]$
Chloride shift

$HHbCO_2$ CO_2

CO_2 *From* <u>CARBAMINO COMPOUND</u>

As *blood passes through capillaries of lungs it takes up approximately 5 to 7 volumes % of OXYGEN. O_2 combines with Haemoglobin (Hb) molecule. It becomes about 95-97% saturated with Oxygen.*

As *blood passes through capillaries of lungs it gives up approximately 4 to 6 volumes % of CARBON DIOXIDE.   A small amount is released from combination with the free amino group in Haemoglobin molecule – Carbamino compound. Most comes from BICARBONATE in R.B.C. and Plasma by processes indicated in diagram.*

NERVOUS CONTROL of RESPIRATORY MOVEMENTS

Normal Respiratory Movements are Involuntary. They are carried out automatically (i.e. without conscious control) through the rhythmical discharge of nerve impulses from <u>CONTROLLING CENTRES in the BRAIN</u>.

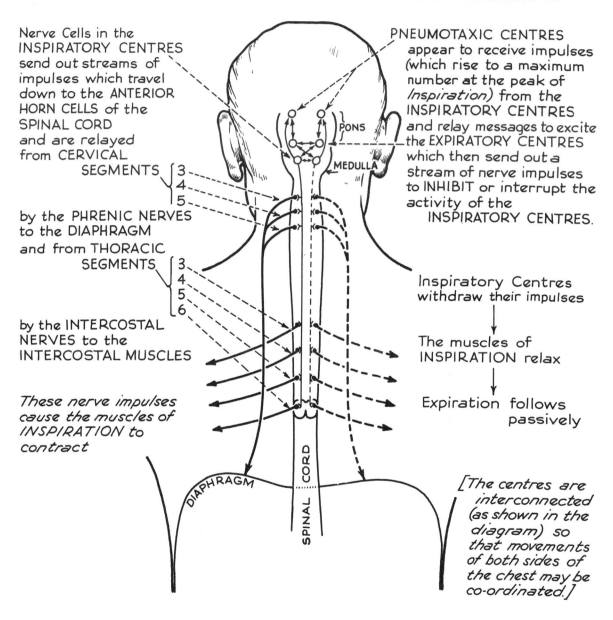

Nerve Cells in the INSPIRATORY CENTRES send out streams of impulses which travel down to the ANTERIOR HORN CELLS of the SPINAL CORD and are relayed from CERVICAL SEGMENTS 3 4 5

by the PHRENIC NERVES to the DIAPHRAGM and from THORACIC SEGMENTS 3 4 5 6

by the INTERCOSTAL NERVES to the INTERCOSTAL MUSCLES

These nerve impulses cause the muscles of INSPIRATION to contract

PNEUMOTAXIC CENTRES appear to receive impulses (which rise to a maximum number at the peak of *Inspiration*) from the INSPIRATORY CENTRES and relay messages to excite the EXPIRATORY CENTRES which then send out a stream of nerve impulses to INHIBIT or interrupt the activity of the INSPIRATORY CENTRES.

PONS

MEDULLA

Inspiratory Centres withdraw their impulses

↓

The muscles of INSPIRATION relax

↓

Expiration follows passively

DIAPHRAGM

SPINAL CORD

[The centres are interconnected (as shown in the diagram) so that movements of both sides of the chest may be co-ordinated.]

CHEMICAL REGULATION of RESPIRATION

The activity of the Respiratory Centres is affected by the chemistry and temperature of the blood supplying them.

The most important factor in regulating the activity of the respiratory centres is the level of CARBON DIOXIDE in the blood passing through them. —— This leads to an alteration in outgoing impulses to respiratory muscles.

Increase in CO_2
Increase in H^+ concentration } stimulates RESPIRATORY CENTRES
(e.g. $CO_2 + H_2O$ in C.S.F. $\rightarrow H_2CO_3 \rightarrow HCO_3^- + H^+$)
— increases rate and depth of breathing.

Fall in blood CO_2 depresses ———— shallow breathing.

Increase in temperature of blood —— quickens but does not deepen respiration.

Cooling of blood ———— slows respiration.

Normally respiration is little affected by slight changes in O_2 content of blood:

Severe lack of O_2 depresses respiratory centre

CHEMO-REFLEXES:
In addition to the direct effect of CO_2 and O_2 on centre

IX

X

Lack of O_2 ————stimulates CHEMORECEPTORS ("oxygen-lack" receptors) in CAROTID BODY and AORTIC BODY ——reflexly stimulates respiration
e.g. as at low atmospheric pressure (high altitude)

NOTE:- These reflexes are usually powerful enough to override the direct depressant action of lack of O_2 on respiratory centres themselves.

The CHEMICAL and NERVOUS means of regulating the activity of RESPIRATORY CENTRES act together to adjust rate and depth of breathing to the needs of the body.

VOLUNTARY and REFLEX FACTORS in the REGULATION of RESPIRATION

Although fundamentally automatic and regulated by chemical factors in the blood, ingoing impulses from many parts of the body also modify the activity of the RESPIRATORY CENTRES and consequently alter the outgoing impulses to the Respiratory muscles to co-ordinate RHYTHM, RATE or DEPTH of breathing with other activities of the body.

Impulses from **HIGHER CENTRES** - PSYCHIC and EMOTIONAL INFLUENCES

{
Voluntary alterations in Breathing.
Interruptions of expiration in SPEECH and SINGING.
Deep inspiration then short spasmodic expirations in LAUGHTER and WEEPING.
Prolonged expiration in SIGHING.
Deep inspiration with mouth open in YAWNING.
Slow shallow breathing in SUSPENSE and CONCENTRATION.
Rapid breathing in FEAR and EXCITEMENT.
}

REFLEX Influences

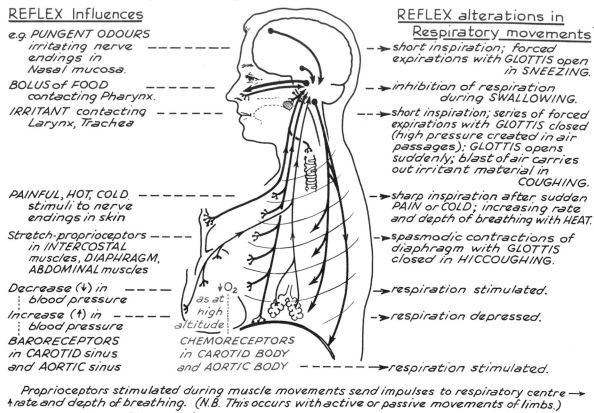

e.g. PUNGENT ODOURS irritating nerve endings in Nasal mucosa. — — — — —

BOLUS of FOOD contacting Pharynx. — — — — —

IRRITANT contacting — — — — — Larynx, Trachea

PAINFUL, HOT, COLD — — — — — — stimuli to nerve endings in skin

Stretch-proprioceptors — — — — in INTERCOSTAL muscles, DIAPHRAGM, ABDOMINAL muscles

Decrease (↓) in — — — — blood pressure

Increase (↑) in — — — — — blood pressure

BARORECEPTORS in CAROTID sinus and AORTIC sinus

↓O_2
as at high altitude

CHEMORECEPTORS in CAROTID BODY and AORTIC BODY — — — — — —

REFLEX alterations in Respiratory movements

→ short inspiration; forced expirations with GLOTTIS open in SNEEZING.

→ inhibition of respiration during SWALLOWING.

— → short inspiration; series of forced expirations with GLOTTIS closed (high pressure created in air passages); GLOTTIS opens suddenly; blast of air carries out irritant material in COUGHING.

→ sharp inspiration after sudden PAIN or COLD; increasing rate and depth of breathing with HEAT.

→ spasmodic contractions of diaphragm with GLOTTIS closed in HICCOUGHING.

→ respiration stimulated.

→ respiration depressed.

→ respiration stimulated.

Proprioceptors stimulated during muscle movements send impulses to respiratory centre → ↑ rate and depth of breathing. (N.B. This occurs with active or passive movements of limbs.)

In normal breathing respiratory rate and rhythm are thought to be influenced rhythmically by the Hering-Breuer Reflex.

| Distension of alveoli at end of inspiration | stimulates stretch receptors in bronchioles | Stream of ingoing impulses passes along vagus nerves to depress inspiratory centres | Withdrawal of outgoing impulses to respiratory muscles → expiration |

CHAPTER 6

EXCRETORY SYSTEM

EXCRETORY SYSTEM

Together with the *Respiratory System* and the *Skin*, the KIDNEYS are the chief *Excretory organs* of the body.

The Kidneys excrete waste products of metabolism and adjust loss of Water and Electrolytes from the body to keep body fluids relatively constant in amount and composition.

To understand the way in which the kidney carries out these functions, it is essential to understand first the way in which it is supplied with blood. About 25% of Left Ventricle's output of blood in each Cardiac Cycle is distributed to kidneys for Filtration.

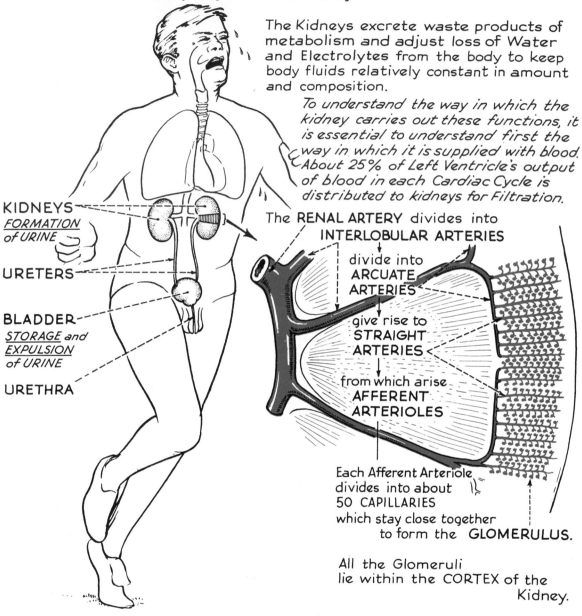

KIDNEYS
FORMATION of URINE

URETERS

BLADDER
STORAGE and EXPULSION of URINE

URETHRA

The **RENAL ARTERY** divides into
INTERLOBULAR ARTERIES

divide into
ARCUATE ARTERIES

give rise to
STRAIGHT ARTERIES

from which arise
AFFERENT ARTERIOLES

Each Afferent Arteriole divides into about
50 CAPILLARIES
which stay close together
to form the **GLOMERULUS.**

All the Glomeruli lie within the CORTEX of the Kidney.

KIDNEY

Each Kidney contains approximately one million microscopic units — *NEPHRONS* — which form URINE.

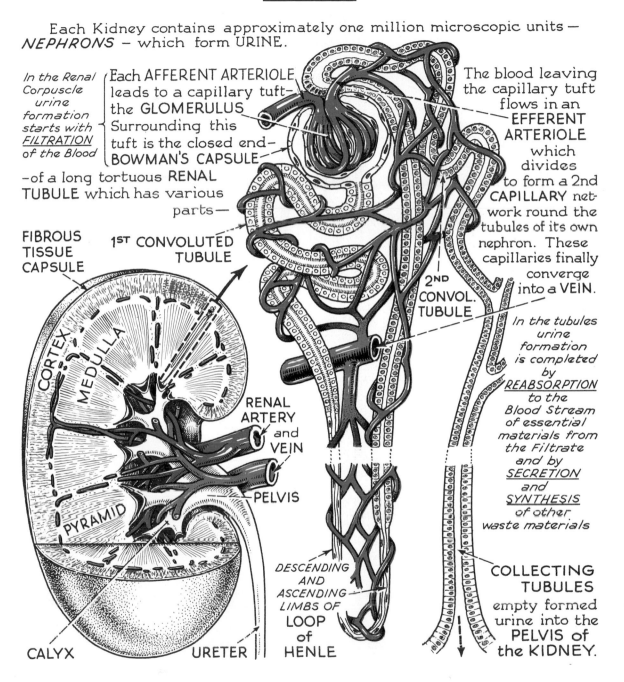

In the Renal Corpuscle urine formation starts with FILTRATION of the Blood

Each AFFERENT ARTERIOLE leads to a capillary tuft— the GLOMERULUS Surrounding this tuft is the closed end— BOWMAN'S CAPSULE

—of a long tortuous RENAL TUBULE which has various parts—

The blood leaving the capillary tuft flows in an EFFERENT ARTERIOLE which divides to form a 2nd CAPILLARY network round the tubules of its own nephron. These capillaries finally converge into a VEIN.

FIBROUS TISSUE CAPSULE

1ST CONVOLUTED TUBULE

2ND CONVOL. TUBULE

In the tubules urine formation is completed by REABSORPTION to the Blood Stream of essential materials from the Filtrate and by SECRETION and SYNTHESIS of other waste materials

CORTEX
MEDULLA

RENAL ARTERY and VEIN

PELVIS

PYRAMID

DESCENDING AND ASCENDING LIMBS OF LOOP of HENLE

COLLECTING TUBULES empty formed urine into the PELVIS of the KIDNEY.

CALYX

URETER

FORMATION of URINE — 1. FILTRATION

About 25% of the left ventricle's total output of blood in each cardiac cycle is distributed through the Renal Arteries to the Kidneys for FILTRATION.

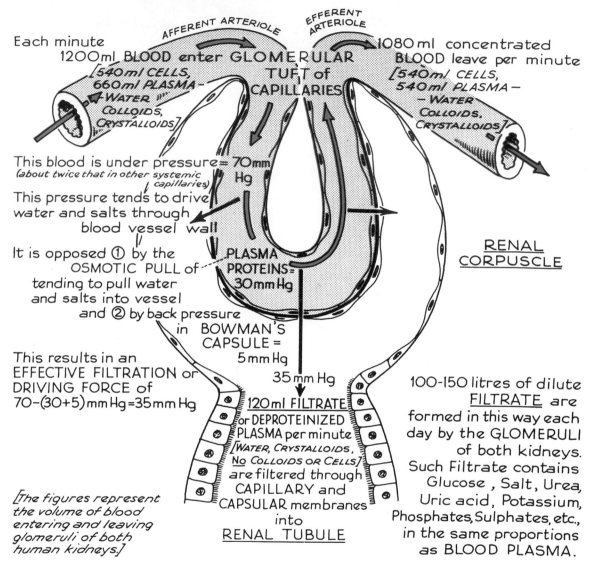

Each minute
1200ml BLOOD enter GLOMERULAR TUFT of CAPILLARIES
[540ml CELLS, 660ml PLASMA — WATER, COLLOIDS, CRYSTALLOIDS]

AFFERENT ARTERIOLE
EFFERENT ARTERIOLE

1080ml concentrated BLOOD leave per minute
[540ml CELLS, 540ml PLASMA — WATER, COLLOIDS, CRYSTALLOIDS]

This blood is under pressure = 70mm Hg (about twice that in other systemic capillaries)

This pressure tends to drive water and salts through blood vessel wall

It is opposed ① by the OSMOTIC PULL of PLASMA PROTEINS = 30mm Hg tending to pull water and salts into vessel and ② by back pressure in BOWMAN'S CAPSULE = 5mm Hg

RENAL CORPUSCLE

This results in an EFFECTIVE FILTRATION or DRIVING FORCE of 70−(30+5)mm Hg = 35mm Hg

35mm Hg

120ml FILTRATE or DEPROTEINIZED PLASMA per minute [WATER, CRYSTALLOIDS, NO COLLOIDS OR CELLS] are filtered through CAPILLARY and CAPSULAR membranes into RENAL TUBULE

[The figures represent the volume of blood entering and leaving glomeruli of both human kidneys.]

100-150 litres of dilute FILTRATE are formed in this way each day by the GLOMERULI of both kidneys. Such Filtrate contains Glucose, Salt, Urea, Uric acid, Potassium, Phosphates, Sulphates, etc., in the same proportions as BLOOD PLASMA.

The Glomerular membrane acts as a simple FILTER —
— i.e. no energy is used up by the cells in filtration.

FORMATION of URINE—2. CONCENTRATION

As it passes along the TUBULE the FILTRATE is CONCENTRATED and essential substances are CONSERVED. The Tubular Epithelium reabsorbs water and selected essential materials into the blood stream.

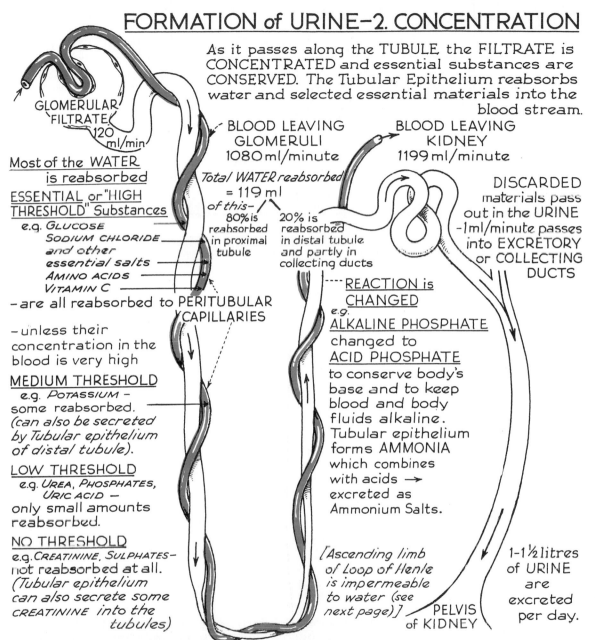

GLOMERULAR FILTRATE 120 ml/min

Most of the WATER is reabsorbed

ESSENTIAL or "HIGH THRESHOLD" Substances

e.g. GLUCOSE
SODIUM CHLORIDE
and other
essential salts
AMINO ACIDS
VITAMIN C

– are all reabsorbed to PERITUBULAR CAPILLARIES

– unless their concentration in the blood is very high

MEDIUM THRESHOLD

e.g. POTASSIUM – some reabsorbed. (can also be secreted by Tubular epithelium of distal tubule).

LOW THRESHOLD

e.g. UREA, PHOSPHATES, URIC ACID — only small amounts reabsorbed.

NO THRESHOLD

e.g. CREATININE, SULPHATES— not reabsorbed at all. (Tubular epithelium can also secrete some CREATININE into the tubules)

BLOOD LEAVING GLOMERULI 1080 ml/minute

Total WATER reabsorbed = 119 ml
of this–
80% is reabsorbed in proximal tubule
20% is reabsorbed in distal tubule and partly in collecting ducts

BLOOD LEAVING KIDNEY 1199 ml/minute

DISCARDED materials pass out in the URINE –1ml/minute passes into EXCRETORY or COLLECTING DUCTS

REACTION is CHANGED

e.g.
ALKALINE PHOSPHATE changed to ACID PHOSPHATE to conserve body's base and to keep blood and body fluids alkaline. Tubular epithelium forms AMMONIA which combines with acids → excreted as Ammonium Salts.

[Ascending limb of Loop of Henle is impermeable to water (see next page)]

PELVIS of KIDNEY

1–1½ litres of URINE are excreted per day.

In concentrating the filtrate and in secretion of certain substances, the tubular epithelial cells use energy, i.e. they do work — much of it against an osmotic gradient.

FORMATION of URINE:
MECHANISM of WATER REABSORPTION

Water is _not_ actively reabsorbed by tubular cells. Its movements are determined _passively_ by the _osmotic gradient_ set up by solutes — chiefly by the _sodium_ salts.

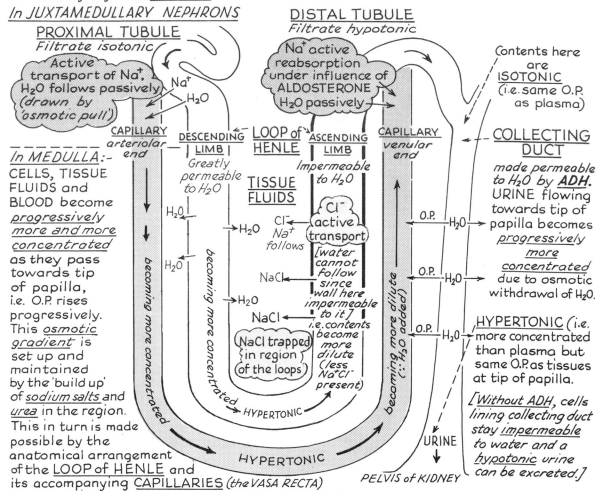

In JUXTAMEDULLARY NEPHRONS

PROXIMAL TUBULE
Filtrate isotonic

Active transport of Na^+. H_2O follows passively _(drawn by 'osmotic pull')_

DISTAL TUBULE
Filtrate hypotonic

Na^+ active reabsorption under influence of ALDOSTERONE H_2O passively

Contents here are ISOTONIC _(i.e. same O.P. as plasma)_

CAPILLARY _arteriolar end_

DESCENDING LIMB
Greatly permeable to H_2O

LOOP of HENLE

ASCENDING LIMB
Impermeable to H_2O

TISSUE FLUIDS

CAPILLARY _venular end_

COLLECTING DUCT
made permeable to H_2O by **ADH**. URINE flowing towards tip of papilla becomes _progressively more concentrated_ due to osmotic withdrawal of H_2O.

In MEDULLA:- CELLS, TISSUE FLUIDS and BLOOD become _progressively more and more concentrated_ as they pass towards tip of papilla, i.e. O.P. rises progressively. This _osmotic gradient_ is set up and maintained by the 'build up' of _sodium salts_ and _urea_ in the region. This in turn is made possible by the anatomical arrangement of the LOOP of HENLE and its accompanying CAPILLARIES _(the VASA RECTA)_

H_2O

H_2O

Cl^- _active transport_

$Cl^- Na^+$ _follows_

NaCl

H_2O

NaCl

NaCl trapped in region of the loops

becoming more concentrated

becoming more concentrated

[water cannot follow since wall here impermeable to it] i.e. contents become more dilute (less Na^+Cl^- present)

HYPERTONIC

HYPERTONIC

becoming more dilute (∵ H_2O added)

O.P. H_2O

O.P. H_2O

O.P. H_2O

HYPERTONIC _(i.e. more concentrated than plasma but same O.P. as tissues at tip of papilla._

[Without ADH, cells lining collecting duct stay **impermeable** to water and a **hypotonic** urine can be excreted.]

URINE

PELVIS of KIDNEY

These form a "COUNTER-CURRENT SYSTEM" (i.e _outflowing_ fluid flows counter to (yet near) _inflowing_). Active transport of Cl with Na ions out of the ascending limb (without accompanying loss of water through its "water-tight" walls) and their resulting diffusion into the descending limb is the key to the chain of events shown.

This mechanism permits maximal [reabsorption of water / concentration of urine] _with minimal expenditure of energy by tubular cells._

The "CLEARANCE" of INULIN in the NEPHRON

The rate of Glomerular Filtration can be measured by injecting a substance which is known to be *filtered off* by the Renal Corpuscle but *not secreted* or *reabsorbed* by the Tubular epithelium.
INULIN is such a substance. It is *filtered* from the Glomerular Capillaries.

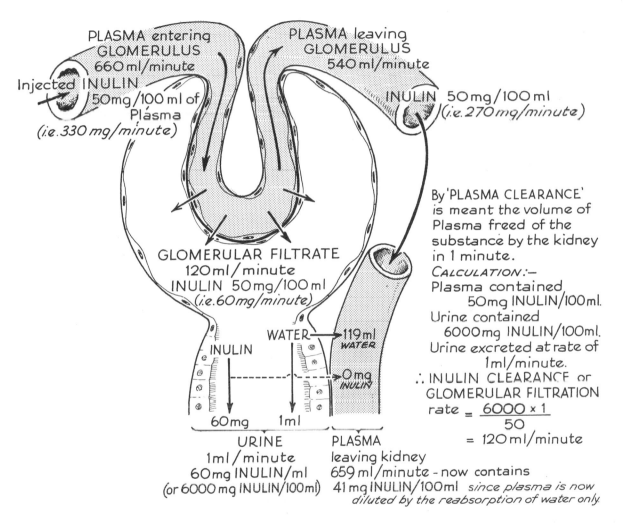

PLASMA entering
GLOMERULUS
660 ml/minute

Injected INULIN
50mg/100 ml of
Plasma
(i.e. 330 mg/minute)

PLASMA leaving
GLOMERULUS
540 ml/minute

INULIN 50mg/100ml
(i.e. 270 mg/minute)

GLOMERULAR FILTRATE
120 ml/minute
INULIN 50mg/100 ml
(i.e. 60mg/minute)

WATER → 119 ml
WATER

INULIN

→ 0mg
INULIN

60mg   1ml

URINE
1ml/minute
60mg INULIN/ml
(or 6000 mg INULIN/100ml)

PLASMA
leaving kidney
659 ml/minute - now contains
41 mg INULIN/100ml *since plasma is now
diluted by the reabsorption of water only.*

By 'PLASMA CLEARANCE'
is meant the volume of
Plasma freed of the
substance by the kidney
in 1 minute.
CALCULATION:–
Plasma contained
50mg INULIN/100ml.
Urine contained
6000mg INULIN/100ml.
Urine excreted at rate of
1ml/minute.
∴ INULIN CLEARANCE or
GLOMERULAR FILTRATION
rate = $\dfrac{6000 \times 1}{50}$
= 120 ml/minute

This idea of clearance can be applied to other substances naturally present such as UREA, or artificially introduced, such as DIODONE.

UREA "CLEARANCE"

UREA, like Inulin, is filtered from the Plasma in the Renal Corpuscle. Unlike Inulin some Urea diffuses back into the Blood Stream from the Tubules.

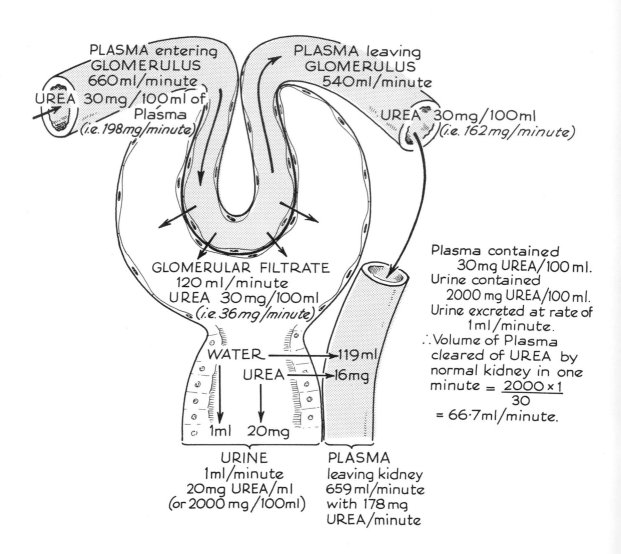

PLASMA entering GLOMERULUS 660ml/minute
UREA 30mg/100ml of Plasma (i.e. 198mg/minute)

PLASMA leaving GLOMERULUS 540ml/minute
UREA 30mg/100ml (i.e. 162mg/minute)

GLOMERULAR FILTRATE 120ml/minute
UREA 30mg/100ml (i.e. 36mg/minute)

WATER → 119ml
UREA → 16mg

1ml 20mg

URINE
1ml/minute
20mg UREA/ml
(or 2000 mg/100ml)

PLASMA
leaving kidney
659 ml/minute
with 178 mg
UREA/minute

Plasma contained 30mg UREA/100 ml. Urine contained 2000 mg UREA/100 ml. Urine excreted at rate of 1ml/minute.
∴ Volume of Plasma cleared of UREA by normal kidney in one minute $= \dfrac{2000 \times 1}{30}$
$= 66.7$ ml/minute.

UREA clearance is used as a test of normal renal function.

140

DIODONE "CLEARANCE"

Certain substances are secreted by the tubules during the formation of urine. Creatinine, which is normally found in urine, is one of these.

Some foreign substances injected into the body are excreted in this way too, e.g. DIODONE *(an iodine-containing material)*.

Such substances can form the basis of tests for checking the efficiency of the Tubular epithelium in any one individual.

Some Diodone is normally *filtered;* some is *excreted* by the tubules.

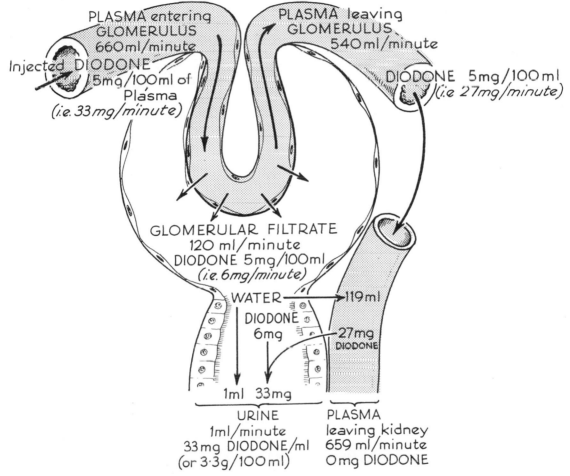

Complete clearance of DIODONE from Plasma in one passage through normal kidney, gauges not only Glomerular Filtrating power but also the efficiency of the Tubular epithelium to secrete.

141

MAINTENANCE of ACID-BASE BALANCE

Large amounts of ACIDS are produced in cells during METABOLISM. In spite of this the pH of the blood and body fluids is kept relatively constant (7.4)

Together with the Respiratory System, the Kidneys are responsible for maintaining this equilibrium.

The ACIDS are first neutralized by BUFFERING AGENTS in the Blood Stream. They cannot be excreted in this form otherwise the body would soon lose its stocks of these ALKALIS.

The Kidney *conserves the body's reserves of Alkali* in several ways :-

1. By *Tubular Secretion of Hydrogen ions* and their *Exchange with Base* in the *Filtrate*.
2. By *Tubular Formation and Secretion of Ammonia* which combines with *Acids* to be excreted as Ammonium Salts

E.g.

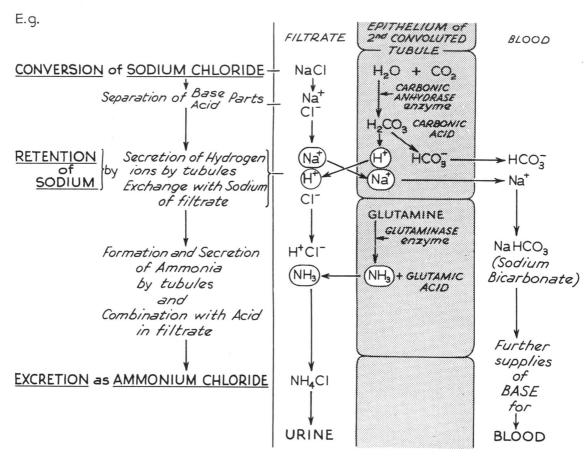

142

MAINTENANCE of ACID-BASE BALANCE

The chief buffering substances of the blood which appear in the filtrate are *Phosphates and Bicarbonates*. The Kidney is believed to *conserve* these as follows:-

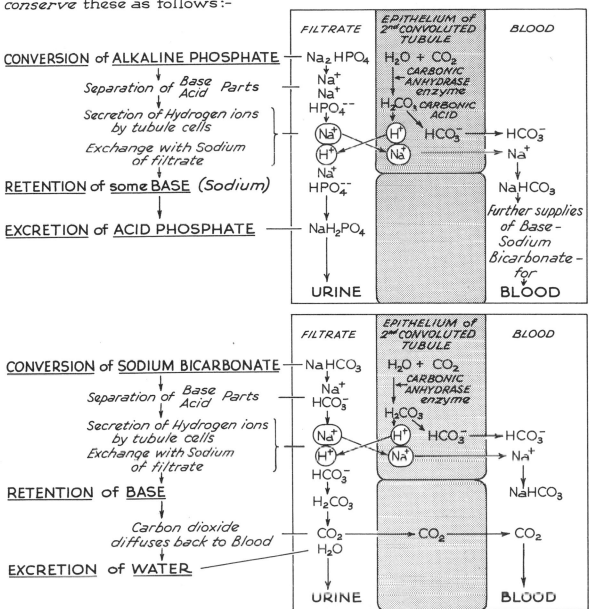

143

REGULATION of WATER BALANCE

As the amounts of water and/or electrolytes in the body fluctuate, excretion of them is adjusted by the *Kidney* so that Body Fluids are restored to normal composition and volume.

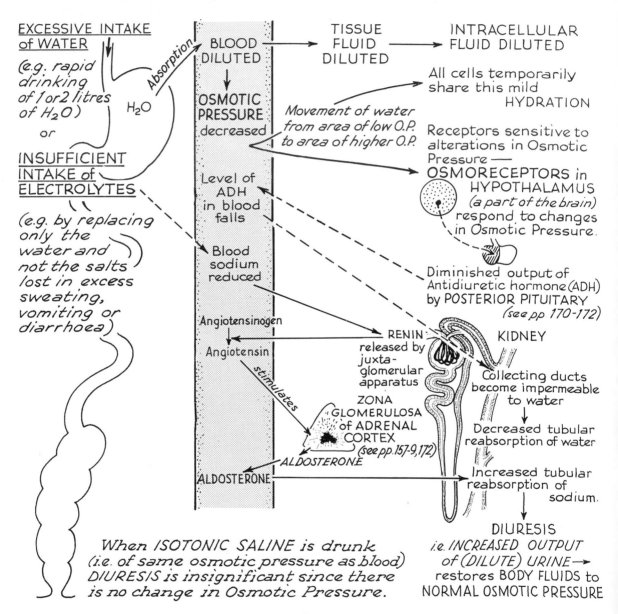

EXCESSIVE INTAKE of WATER

(e.g. rapid drinking of 1 or 2 litres of H_2O)

or

INSUFFICIENT INTAKE of ELECTROLYTES

(e.g. by replacing only the water and not the salts lost in excess sweating, vomiting or diarrhoea)

H_2O

Absorption

BLOOD DILUTED

↓

OSMOTIC PRESSURE decreased

TISSUE FLUID DILUTED

INTRACELLULAR FLUID DILUTED

All cells temporarily share this mild HYDRATION

Movement of water from area of low O.P. to area of higher O.P.

Receptors sensitive to alterations in Osmotic Pressure —

OSMORECEPTORS in HYPOTHALAMUS *(a part of the brain)* respond to changes in Osmotic Pressure.

Level of ADH in blood falls

Diminished output of Antidiuretic hormone (ADH) by POSTERIOR PITUITARY *(see pp 170-172)*

Blood sodium reduced

Angiotensinogen

↓

Angiotensin

stimulates

RENIN released by juxta-glomerular apparatus

KIDNEY

Collecting ducts become impermeable to water

↓

Decreased tubular reabsorption of water

ZONA GLOMERULOSA of ADRENAL CORTEX *(see pp.157-9,172)*

ALDOSTERONE

ALDOSTERONE

Increased tubular reabsorption of sodium.

DIURESIS *i.e. INCREASED OUTPUT OF (DILUTE) URINE →* restores BODY FLUIDS to NORMAL OSMOTIC PRESSURE

When *ISOTONIC SALINE is drunk (i.e. of same osmotic pressure as blood) DIURESIS is insignificant since there is no change in Osmotic Pressure.*

REGULATION of WATER BALANCE

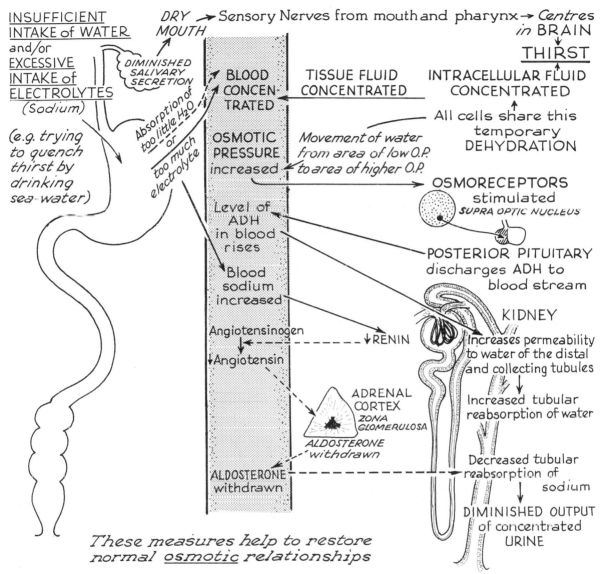

INSUFFICIENT INTAKE of WATER and/or EXCESSIVE INTAKE of ELECTROLYTES (Sodium)

(e.g. trying to quench thirst by drinking sea-water)

DRY MOUTH → Sensory Nerves from mouth and pharynx → Centres in BRAIN

THIRST

DIMINISHED SALIVARY SECRETION

Absorption of too little H₂O or too much electrolyte

BLOOD CONCENTRATED

TISSUE FLUID CONCENTRATED

INTRACELLULAR FLUID CONCENTRATED

All cells share this temporary DEHYDRATION

OSMOTIC PRESSURE increased

Movement of water from area of low O.P. to area of higher O.P.

OSMORECEPTORS stimulated
SUPRA OPTIC NUCLEUS

Level of ADH in blood rises

POSTERIOR PITUITARY discharges ADH to blood stream

Blood sodium increased

KIDNEY

Angiotensinogen → ↓RENIN

↓Angiotensin

Increases permeability to water of the distal and collecting tubules

Increased tubular reabsorption of water

ADRENAL CORTEX
ZONA GLOMERULOSA

ALDOSTERONE withdrawn

Decreased tubular reabsorption of sodium

ALDOSTERONE withdrawn

DIMINISHED OUTPUT of concentrated URINE

These measures help to restore normal osmotic relationships

In the maintenance of water balance, loss (or conservation) of water and of electrolytes are closely linked. Another example, shown on page 172, deals with ALDOSTERONE, its rôle in governing reabsorption of sodium by kidney tubules, and its tie-up with ADH in maintaining the normal volume of extracellular fluids including blood.

URINARY BLADDER and URETERS

A resistant, distensible *Transitional Epithelium* lines all *Urinary Passages*.

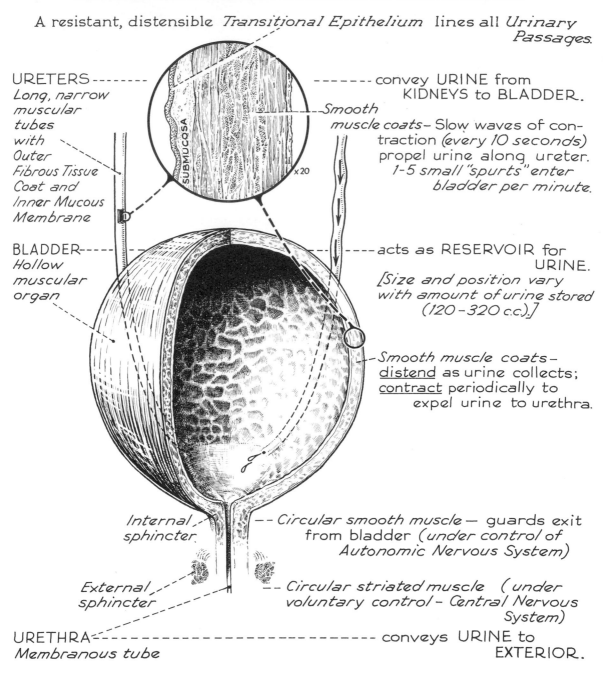

URETERS ------- *Long, narrow muscular tubes with Outer Fibrous Tissue Coat and Inner Mucous Membrane*

------ convey URINE from KIDNEYS to BLADDER.

Smooth muscle coats– Slow waves of contraction (*every 10 seconds*) propel urine along ureter. *1-5 small "spurts" enter bladder per minute.*

SUBMUCOSA

×20

BLADDER---- *Hollow muscular organ*

---- acts as RESERVOIR for URINE. *[Size and position vary with amount of urine stored (120-320 c.c.).]*

Smooth muscle coats– <u>distend</u> as urine collects; <u>contract</u> periodically to expel urine to urethra.

Internal sphincter

--Circular smooth muscle — guards exit from bladder (*under control of Autonomic Nervous System*)

External sphincter

-- Circular striated muscle (*under voluntary control – Central Nervous System*)

URETHRA------ *Membranous tube*

------- conveys URINE to EXTERIOR.

146

STORAGE and EXPULSION of URINE

URINE is formed continuously by the KIDNEYS.   It collects, drop by drop, in the URINARY BLADDER which expands to hold about 300 ml. When the Bladder is full the desire to *void urine* is experienced

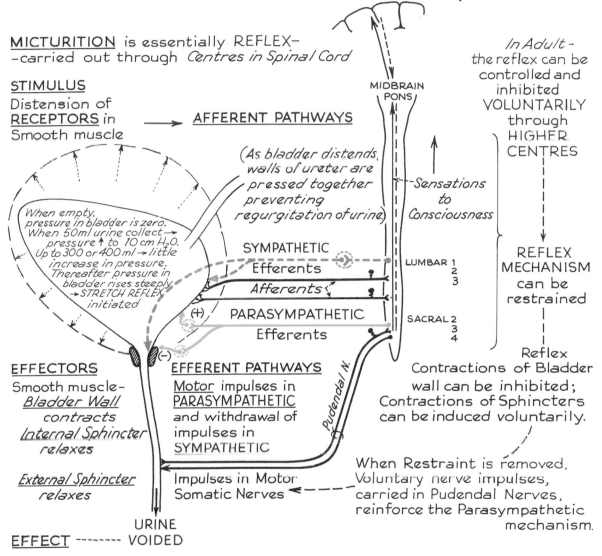

MICTURITION is essentially REFLEX—
—carried out through *Centres in Spinal Cord*

*In Adult —
the reflex can be
controlled and
inhibited
VOLUNTARILY
through
HIGHER
CENTRES*

MIDBRAIN
PONS

STIMULUS
Distension of
RECEPTORS in
Smooth muscle

AFFERENT PATHWAYS

*(As bladder distends,
walls of ureter are
pressed together
preventing
regurgitation of urine.)*

—Sensations
to
Consciousness

*When empty,
pressure in bladder is zero.
When 50 ml urine collect →
pressure ↑ to 10 cm H₂O.
Up to 300 or 400 ml → little
increase in pressure.
Thereafter pressure in
bladder rises steeply
→ STRETCH REFLEX
initiated*

SYMPATHETIC
Efferents

Afferents

LUMBAR 1
2
3

REFLEX
MECHANISM
can be
restrained

(+) PARASYMPATHETIC
Efferents

SACRAL 2
3
4

(-)

Reflex
Contractions of Bladder
wall can be inhibited;
Contractions of Sphincters
can be induced voluntarily.

EFFECTORS
Smooth muscle—
Bladder Wall
contracts
Internal Sphincter
relaxes

External Sphincter
relaxes

EFFERENT PATHWAYS
Motor impulses in
PARASYMPATHETIC
and withdrawal of
impulses in
SYMPATHETIC

Pudendal N.

Impulses in Motor
Somatic Nerves

When Restraint is removed,
Voluntary nerve impulses,
carried in Pudendal Nerves,
reinforce the Parasympathetic
mechanism.

URINE

EFFECT -------- VOIDED

*When Bladder is empty and beginning to fill —
Inhibition of Parasympathetic* } *Relaxation of Bladder Wall.
Activation of Sympathetic* } *Constriction of Sphincters.*

URINE

VOLUME: *In Adult*
 1000 - 1500 ml/24 hours

SPECIFIC GRAVITY 1·001 - 1·040

} Vary with *Fluid Intake* and with *Fluid Output* from other routes — *Skin, Lungs, Gut*.
[Volume reduced during *Sleep* and *Muscular Exercise* :
 Specific Gravity greater on Protein
 diet.]

REACTION Normally slightly acid
 (pH around 6) — *Varies* with *Diet*
[acid on ordinary mixed diet:
 alkaline on vegetarian diet].

COLOUR

YELLOW
due to
UROCHROME
pigment-
probably from
destruction of
tissue
 protein.

More
concentrated
and DARKER
in early
morning —
less water
excreted at
night but
unchanged
amounts of
Urinary
 Solids.

ODOUR

AROMATIC
when fresh→
AMMONIACAL
on standing
due to
bacterial
decomposition
of UREA to
AMMONIA.

COMPOSITION

Grams excreted in 24 hours

WATER 96% 1000–1500

INORGANIC SUBSTANCES

Sodium	6
Chloride	7
Calcium	0·2
Potassium	2
Phosphates	1·7
Sulphates	1·8

[These figures are approximate and vary widely in healthy individuals]

ORGANIC SUBSTANCES

Urea 2% 20 - 30 — derived from breakdown of *Protein* — therefore varies with Protein in Diet.

Uric Acid 0·6 — comes from *Purine* of Food and Body Tissues.

Creatinine 1·2 — from breakdown of Body Tissues; uninfluenced by amount of dietary Protein.

Ammonia 0·5-0·9 — formed in Kidney from *Glutamine* brought to it by Blood Stream; varies with amounts of acid substances requiring neutralization in the Kidney.

[In the Newborn, Volume and Specific Gravity are low and Composition varies.]

CHAPTER 7

ENDOCRINE SYSTEM

ENDOCRINE SYSTEM

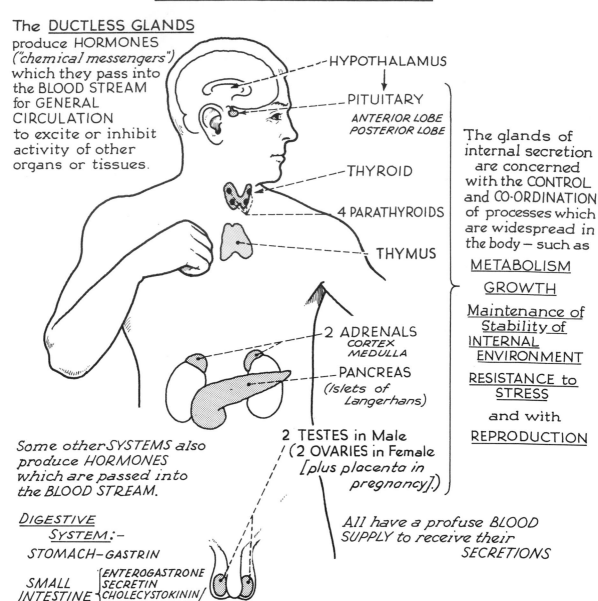

The <u>DUCTLESS GLANDS</u>
produce HORMONES
("chemical messengers")
which they pass into
the BLOOD STREAM
for GENERAL
CIRCULATION
to excite or inhibit
activity of other
organs or tissues.

--- HYPOTHALAMUS

--- PITUITARY
ANTERIOR LOBE
POSTERIOR LOBE

--- THYROID

--- 4 PARATHYROIDS

--- THYMUS

2 ADRENALS
CORTEX
MEDULLA

--- PANCREAS
(Islets of
Langerhans)

2 TESTES in Male
(2 OVARIES in Female
[plus placenta in
pregnancy].)

The glands of
internal secretion
are concerned
with the CONTROL
and CO-ORDINATION
of processes which
are widespread in
the body – such as

<u>METABOLISM</u>

<u>GROWTH</u>

<u>Maintenance of
Stability of
INTERNAL
ENVIRONMENT</u>

<u>RESISTANCE to
STRESS</u>

and with

<u>REPRODUCTION</u>

All have a profuse BLOOD
SUPPLY to receive their
SECRETIONS

*Some other SYSTEMS also
produce HORMONES
which are passed into
the BLOOD STREAM.*

<u>DIGESTIVE
SYSTEM</u>:–
STOMACH – GASTRIN

SMALL
INTESTINE
{ ENTEROGASTRONE
SECRETIN
CHOLECYSTOKININ/
PANCREOZYMIN

KIDNEY
{ e.g. ERYTHROPOIETIN
RENIN
1:25 DHCC

Although each ENDOCRINE GLAND has
specific functions all are interdependent.
Overactivity or Underactivity of one
tends to affect the whole system.

THYROID

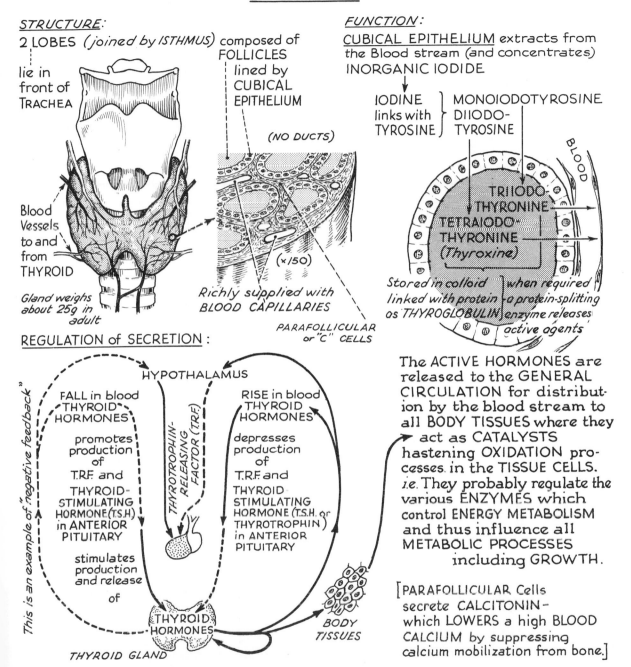

STRUCTURE:

2 LOBES (joined by ISTHMUS) composed of FOLLICLES lined by CUBICAL EPITHELIUM

lie in front of TRACHEA

(NO DUCTS)

Blood Vessels to and from THYROID

Gland weighs about 25g in adult

(×/50)

Richly supplied with BLOOD CAPILLARIES

PARAFOLLICULAR or "C" CELLS

FUNCTION:

CUBICAL EPITHELIUM extracts from the Blood stream (and concentrates) INORGANIC IODIDE

IODINE links with TYROSINE ⎱ MONOIODOTYROSINE DIIODO-TYROSINE

BLOOD

TRIIODO-THYRONINE
TETRAIODO-THYRONINE (Thyroxine)

Stored in colloid linked with protein as THYROGLOBULIN — when required a protein-splitting enzyme releases active agents

REGULATION of SECRETION:

This is an example of "negative feedback"

HYPOTHALAMUS

FALL in blood THYROID HORMONES promotes production of T.R.F. and THYROID-STIMULATING HORMONE (T.S.H.) in ANTERIOR PITUITARY stimulates production and release of

THYROTROPHIN-RELEASING FACTOR (T.R.F.)

RISE in blood THYROID HORMONES depresses production of T.R.F. and THYROID-STIMULATING HORMONE (T.S.H. or THYROTROPHIN) in ANTERIOR PITUITARY

THYROID HORMONES

THYROID GLAND

BODY TISSUES

The ACTIVE HORMONES are released to the GENERAL CIRCULATION for distribution by the blood stream to all BODY TISSUES where they act as CATALYSTS hastening OXIDATION processes in the TISSUE CELLS. i.e. They probably regulate the various ENZYMES which control ENERGY METABOLISM and thus influence all METABOLIC PROCESSES including GROWTH.

[PARAFOLLICULAR Cells secrete CALCITONIN — which LOWERS a high BLOOD CALCIUM by suppressing calcium mobilization from bone.]

UNDERACTIVITY of THYROID

If the THYROID shows *atrophy* of its secretory cells or is inadequately stimulated by the Anterior Pituitary:-

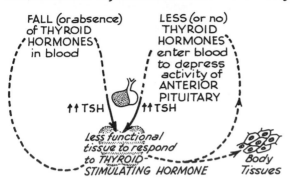

FALL (or absence) of THYROID HORMONES in blood

LESS (or no) THYROID HORMONES enter blood to depress activity of ANTERIOR PITUITARY

↑↑TSH ↑↑TSH

Less functional tissue to respond to THYROID-STIMULATING HORMONE

Body Tissues

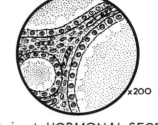

×200

Insufficient HORMONAL SECRETION released to Blood Stream.

TISSUE OXIDATIONS are depressed, *i.e.* Rate at which cells use energy is reduced.

The Basal Metabolic Rate falls.

Less Heat is produced.

Body Temperature falls (and person feels COLD).

Energy units are stored with water.

SKIN - Thick, leathery, puffy, yellow (due to circulating carotene).

In the *ADULT*

MYXOEDEMA

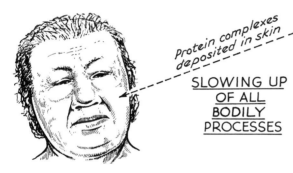

protein complexes deposited in skin

SLOWING UP OF ALL BODILY PROCESSES

Blood cholesterol increases.
Appetite is reduced; Weight increases.
Gut movements sluggish → Constipation.
Heart and Respiratory Rates and Blood Pressure reduced.
Thought processes slow down →
Lethargy; Apathy; Somnolence.
HAIR – Brittle, sparse, dry.
Slow, husky voice. Bone marrow suppressed → ANAEMIA.

In the *CHILD* —— e.g. congenital absence of the gland →

CRETIN

GROSS ↓ DWARFING

FAILURE of SKELETAL SEXUAL MENTAL } GROWTH and DEVELOPMENT

All "milestones" of babyhood are delayed.

THYROXINE (taken by mouth) restores individuals to normal.

OVERACTIVITY of THYROID

Overactivity (e.g. tumour) or the presence of an abnormal long-acting thyroid stimulating hormone (L.A.T.S.) from the anterior pituitary.

Great RISE in blood THYROID HORMONES

Great FALL in blood THYROID HORMONES

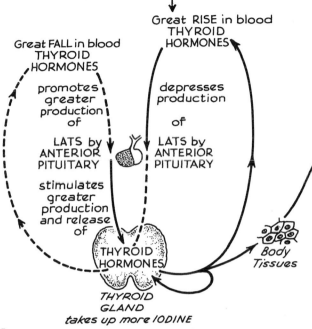

promotes greater production of

depresses production of

LATS by ANTERIOR PITUITARY

LATS by ANTERIOR PITUITARY

stimulates greater production and release of

THYROID HORMONES

Body Tissues

THYROID GLAND
takes up more IODINE

[*N.B. If the excess thyroxine is formed by tumour tissue this is outwith the negative feedback control of TSH. LATS is not suppressed by ↑Thyroxine.*]

×200

EXCESS THYROID HORMONES are distributed by blood stream to the Tissues of the Body. ⟶
SPEED up OXIDATIONS in the cells, i.e. rate at which all cells use ENERGY.

The Basal Metabolic Rate is raised. As a by-product of this increased cellular activity more heat is produced ⟶ rise in Body Temperature (*person feels WARM*).
*Skin hot and flushed.
          Profuse Sweating.
Energy stores of body (i.e. GLYGOGEN and FAT) are depleted.

SPEEDING UP
OF ALL
BODILY
PROCESSES

{
Appetite increases but weight falls.
Movements of digestive tract are increased ⟶ Diarrhoea.
*Heart and Respiratory Rates rise.
*Blood Pressure is raised.
Muscular tremor and nervousness are marked.
Person becomes excitable, irritable and apprehensive.
}

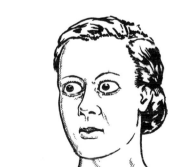

(*Some of these effects are due to activation of hypothalamus and sympathetic nervous system by ↑HEAT)

[EXOPHTHALMOS (*protrusion of eyeballs*) may be due to an excess of the abnormal L.A.T.S. It is not due to an excess of Thyroid Hormones.]

Surgical removal of part of the overactive gland reduces the Thyroid activity.

PARATHYROIDS

FOUR small glands composed of *cords of cells* which secrete Parathyroid Hormone – **PARATHORMONE**
or PTH

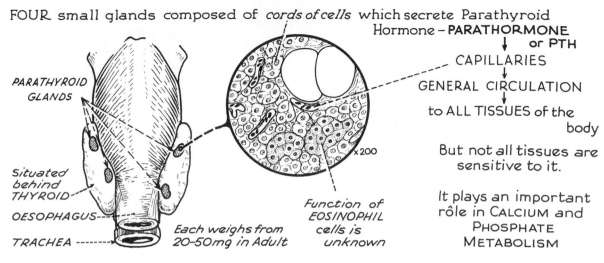

PARATHYROID GLANDS

Situated behind THYROID

OESOPHAGUS

TRACHEA

Each weighs from 20-50mg in Adult

×200

Function of EOSINOPHIL cells is unknown

CAPILLARIES

GENERAL CIRCULATION

to ALL TISSUES of the body

But not all tissues are sensitive to it.

It plays an important rôle in CALCIUM and PHOSPHATE METABOLISM

PARATHORMONE acts on <u>KIDNEY TUBULES, BONE</u> and on <u>GUT</u> to maintain ionized BLOOD CALCIUM level at 11mg/100ml PLASMA *(necessary for normal neuromuscular excitability)*.

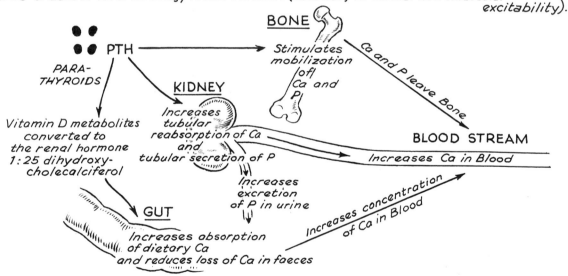

PARA-THYROIDS

PTH

BONE

Stimulates mobilization of Ca and P

KIDNEY

Increases tubular reabsorption of Ca and tubular secretion of P

Ca and P leave Bone

BLOOD STREAM

Increases Ca in Blood

Vitamin D metabolites converted to the renal hormone 1:25 dihydroxy-cholecalciferol

Increases excretion of P in urine

Increases concentration of Ca in Blood

GUT

Increases absorption of dietary Ca and reduces loss of Ca in faeces

Alterations in concentration of CALCIUM ions in extracellular fluids control Parathyroid activity. A rise in blood Calcium depresses Parathyroid secretion. A fall in Calcium increases Parathyroid secretion. [Note: This response to the level of plasma Ca differs from calcitonin's — see "thyroid."]

154

UNDERACTIVITY of PARATHYROIDS

Atrophy or removal of Parathyroid tissue causes a fall in BLOOD CALCIUM level and increased excitability of Neuromuscular tissue. This leads to severe convulsive disorder — TETANY.

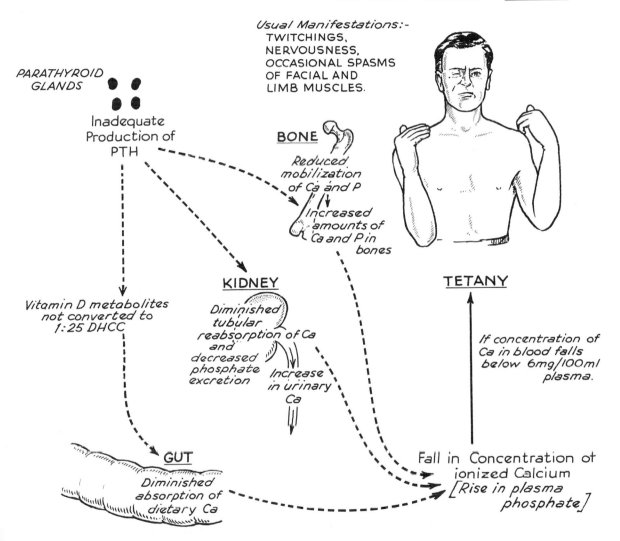

Usual Manifestations:-
TWITCHINGS,
NERVOUSNESS,
OCCASIONAL SPASMS
OF FACIAL AND
LIMB MUSCLES.

PARATHYROID GLANDS

Inadequate Production of PTH

BONE
Reduced mobilization of Ca and P
Increased amounts of Ca and P in bones

TETANY

Vitamin D metabolites not converted to 1:25 DHCC

KIDNEY
Diminished tubular reabsorption of Ca and decreased phosphate excretion
Increase in urinary Ca

If concentration of Ca in blood falls below 6mg/100ml plasma.

GUT
Diminished absorption of dietary Ca

Fall in Concentration of ionized Calcium
[Rise in plasma phosphate]

[Note the inverse relationship between plasma calcium and inorganic phosphate]

Symptoms are relieved by injection of Calcium, large doses of a Vit.D compound and Parathormone.

OVERACTIVITY of PARATHYROIDS

Overactivity of the Parathyroids *(due often to tumour)* leads to rise in BLOOD CALCIUM level and eventually to <u>OSTEITIS FIBROSA CYSTICA</u>.

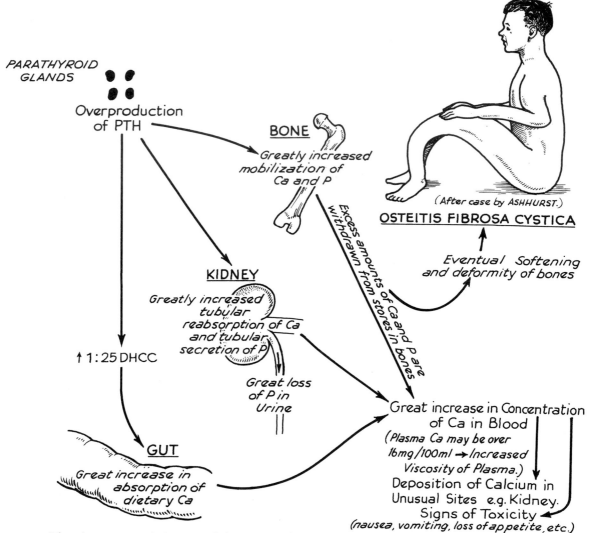

PARATHYROID GLANDS

Overproduction of PTH

BONE
Greatly increased mobilization of Ca and P

↑1:25 DHCC

KIDNEY
Greatly increased tubular reabsorption of Ca and tubular secretion of P

Great loss of P in Urine

Excess amounts of Ca and P are withdrawn from stores in bones

GUT
Great increase in absorption of dietary Ca

(After case by ASHHURST.)
OSTEITIS FIBROSA CYSTICA

Eventual Softening and deformity of bones

Great increase in Concentration of Ca in Blood
(Plasma Ca may be over 16mg/100ml → Increased Viscosity of Plasma.)
Deposition of Calcium in Unusual Sites e.g. Kidney.
Signs of Toxicity
(nausea, vomiting, loss of appetite, etc.)

The increased level of blood calcium eventually leads to excessive loss of CALCIUM in URINE (in spite of ↑reabsorption) and also of WATER since the salts are excreted in solution. POLYURIA and THIRST result.

Excision of the overactive Parathyroid tissue abolishes syndrome.

ADRENAL GLANDS

There are TWO Adrenal Glands.
They lie close to the kidneys.

*Each has an outer CORTEX
and an inner
MEDULLA*

The Adrenal Cortex secretes
steroid HORMONES
into the BLOOD STREAM.
These travel to all tissues of
the body.
The Adrenal Steroids
share the same functions
but to varying degrees:—

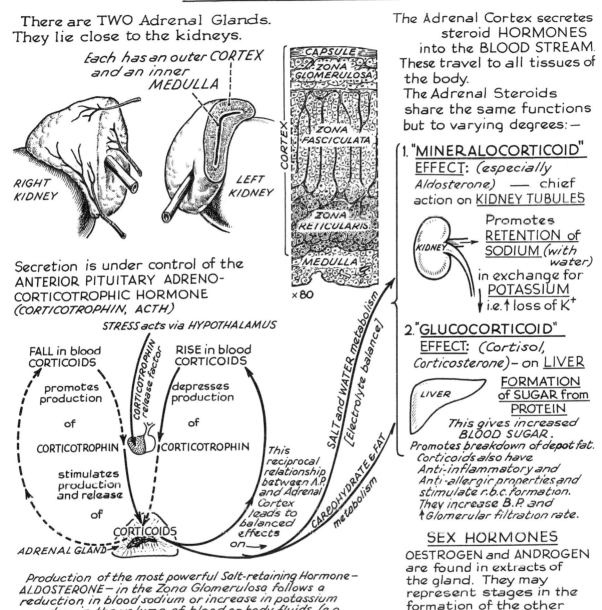

RIGHT
KIDNEY

LEFT
KIDNEY

CAPSULE
ZONA
GLOMERULOSA

ZONA
FASCICULATA

ZONA
RETICULARIS

MEDULLA

CORTEX

×80

Secretion is under control of the
ANTERIOR PITUITARY ADRENO-
CORTICOTROPHIC HORMONE
(CORTICOTROPHIN, ACTH)

STRESS acts via HYPOTHALAMUS

FALL in blood
CORTICOIDS

promotes
production

of

CORTICOTROPHIN

stimulates
production
and release

of

CORTICOTROPHIN release factor

RISE in blood
CORTICOIDS

depresses
production

of

CORTICOTROPHIN

*This
reciprocal
relationship
between A.P.
and Adrenal
Cortex
leads to
balanced
effects
on* →

CORTICOIDS

ADRENAL GLAND

*SALT and WATER metabolism
[Electrolyte balance]*

*CARBOHYDRATE & FAT
metabolism*

1. "MINERALOCORTICOID"
EFFECT: *(especially
Aldosterone)* — chief
action on KIDNEY TUBULES

KIDNEY

Promotes
RETENTION of
SODIUM *(with
water)*
in exchange for
POTASSIUM
i.e. ↑ loss of K⁺

2. "GLUCOCORTICOID"
EFFECT: *(Cortisol,
Corticosterone)* – on LIVER

LIVER

FORMATION
of SUGAR from
PROTEIN
*This gives increased
BLOOD SUGAR.
Promotes breakdown of depot fat.
Corticoids also have
Anti-inflammatory and
Anti-allergic properties and
stimulate r.b.c. formation.
They increase B.P. and
↑ Glomerular filtration rate.*

SEX HORMONES

OESTROGEN and ANDROGEN
are found in extracts of
the gland. They may
represent stages in the
formation of the other
corticoids

*Production of the most powerful Salt-retaining Hormone—
ALDOSTERONE — in the Zona Glomerulosa follows a
reduction in blood sodium or increase in potassium
or a drop in the volume of blood or body fluids (e.g.
after haemorrhage, excess vomiting or diarrhoea).*

The Adrenal Cortex is <u>essential to life</u> and plays an important rôle in states of stress.

UNDERACTIVITY of ADRENAL CORTEX

Atrophy of Adrenal Cortex *(occasionally occurs with destructive disease*
gives *of the gland, e.g. Tuberculosis)*
Inadequate Production of all CORTICOIDS :-

↓"MINERALOCORTICOID" EFFECT :

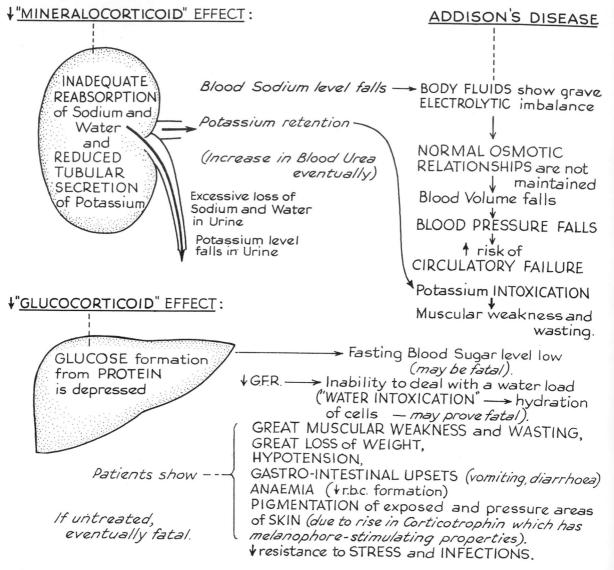

ADDISON'S DISEASE

INADEQUATE REABSORPTION of Sodium and Water and REDUCED TUBULAR SECRETION of Potassium

Blood Sodium level falls ⟶ BODY FLUIDS show grave ELECTROLYTIC imbalance

Potassium retention

(Increase in Blood Urea eventually)

Excessive loss of Sodium and Water in Urine

Potassium level falls in Urine

NORMAL OSMOTIC RELATIONSHIPS are not maintained
Blood Volume falls

BLOOD PRESSURE FALLS

↑ risk of CIRCULATORY FAILURE

Potassium INTOXICATION

Muscular weakness and wasting.

↓"GLUCOCORTICOID" EFFECT :

GLUCOSE formation from PROTEIN is depressed ⟶ Fasting Blood Sugar level low *(may be fatal).*

↓G.F.R. ⟶ Inability to deal with a water load ("WATER INTOXICATION" ⟶ hydration of cells — *may prove fatal).*

Patients show -- - {
GREAT MUSCULAR WEAKNESS and WASTING,
GREAT LOSS of WEIGHT,
HYPOTENSION,
GASTRO-INTESTINAL UPSETS *(vomiting, diarrhoea)*
ANAEMIA (↓r.b.c. formation)
PIGMENTATION of exposed and pressure areas of SKIN *(due to rise in Corticotrophin which has melanophore-stimulating properties).*
↓resistance to STRESS and INFECTIONS.
}

If untreated, eventually fatal.

Administration of Adrenal Hormones and Sodium Chloride restores individual to normal.

OVERACTIVITY of ADRENAL CORTEX

Overactivity or tumour of Adrenal Cortex
        may give
Overproduction of any or all of the CORTICOIDS :-

e.g. __ALDOSTERONE__ ⟶ PRIMARY ALDOSTERONISM

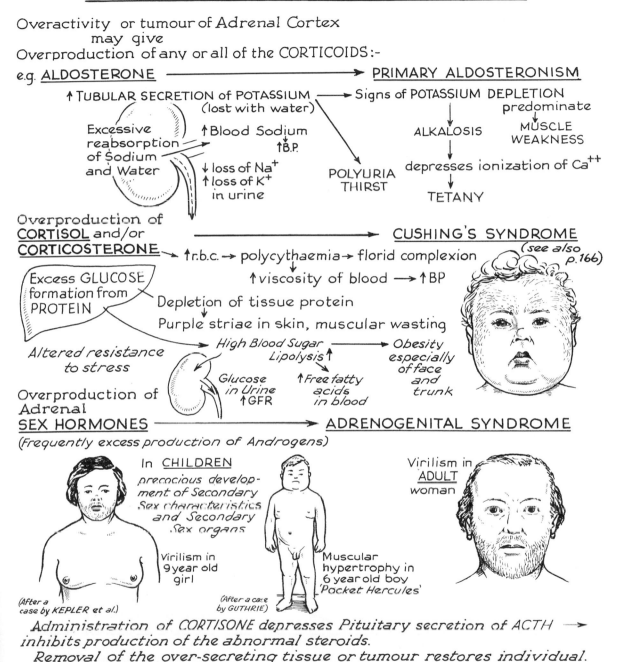

↑TUBULAR SECRETION OF POTASSIUM ⟶ Signs of POTASSIUM DEPLETION
(lost with water)   predominate

Excessive   ↑Blood Sodium   ALKALOSIS   MUSCLE
reabsorption   ↑B.P.   WEAKNESS
of Sodium
and Water   ↓ loss of Na^+ depresses ionization of Ca^{++}
                   ↑ loss of K^+ POLYURIA
                   in urine   THIRST   TETANY

Overproduction of
__CORTISOL__ and/or ⟶ CUSHING'S SYNDROME
__CORTICOSTERONE__ ↘ ↑r.b.c. ⟶ polycythaemia ⟶ florid complexion   *(see also p.166)*

                   ↑viscosity of blood ⟶ ↑BP

Excess GLUCOSE
formation from   Depletion of tissue protein
PROTEIN
                   Purple striae in skin, muscular wasting

Altered resistance   *High Blood Sugar* ⟶ *Obesity*
to stress   *Lipolysis↑*   *especially*
                                                  of face
                  Glucose   *↑Free fatty*   *and*
Overproduction of *in Urine*   *acids*   *trunk*
Adrenal   *↑GFR*   *in blood*
__SEX HORMONES__ ⟶ ADRENOGENITAL SYNDROME

(Frequently excess production of Androgens)

In __CHILDREN__   Virilism in
precocious develop-   __ADULT__
ment of Secondary   woman
Sex characteristics
and Secondary
Sex organs

Virilism in   Muscular
9 year old   hypertrophy in
girl   6 year old boy
                         'Pocket Hercules'

(After a case by KEPLER et al.)   *(After a case by GUTHRIE)*

Administration of CORTISONE depresses Pituitary secretion of ACTH ⟶
inhibits production of the abnormal steroids.
Removal of the over-secreting tissue or tumour restores individual.

159

ADRENAL MEDULLA

The Adrenal Medulla arises from the same primitive tissue as the Ganglion cells of the Sympathetic Nervous System.

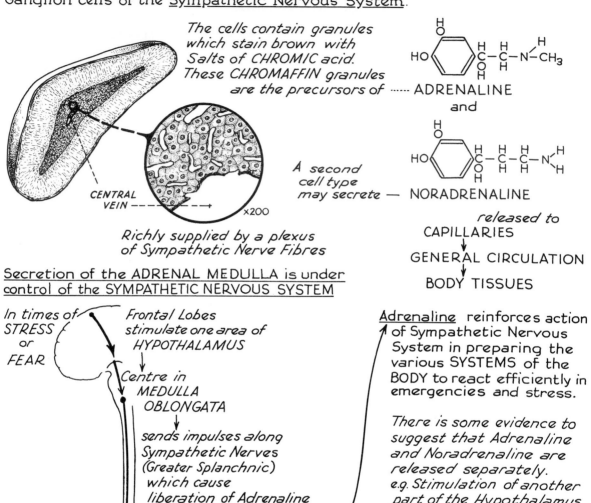

The cells contain granules which stain brown with Salts of CHROMIC acid. These CHROMAFFIN granules are the precursors of ······ ADRENALINE

and

A second cell type may secrete — NORADRENALINE

released to
CAPILLARIES
↓
GENERAL CIRCULATION
↓
BODY TISSUES

CENTRAL VEIN

×200

Richly supplied by a plexus of Sympathetic Nerve Fibres

Secretion of the ADRENAL MEDULLA is under control of the SYMPATHETIC NERVOUS SYSTEM

In times of STRESS or FEAR

Frontal Lobes stimulate one area of HYPOTHALAMUS

Centre in MEDULLA OBLONGATA

sends impulses along Sympathetic Nerves (Greater Splanchnic) which cause liberation of Adrenaline by Adrenal Medullae into Blood Stream

O_2 lack in Blood Stream has a direct effect on Adrenal Medullae

↓O_2

Adrenaline reinforces action of Sympathetic Nervous System in preparing the various SYSTEMS of the BODY to react efficiently in emergencies and stress.

There is some evidence to suggest that Adrenaline and Noradrenaline are released separately. e.g. Stimulation of another part of the Hypothalamus apparently leads to release of Noradrenaline into Blood Stream
↓
general vasoconstriction (except of coronaries)
↓
Rise in Blood Pressure.

ADRENALINE

Under quiet resting conditions the blood contains very little adrenaline. During excitement or circumstances which demand special efforts adrenaline is released into the blood stream, and is responsible for the following actions summed up as the "FIGHT or FLIGHT" FUNCTION of the Adrenal Medullae.

It _Constricts_ Smooth Muscle of Skin →
      Hairs 'stand on end'; 'Gooseflesh'.

Dilates Pupil of Eye to
      admit more light.

Constricts Smooth Muscle of Abdominal Blood Vessels and Cutaneous Blood Vessels → Pallor with Fright.

Dilates Smooth Muscle in arterioles of
      Skeletal Muscles.

Excites Cardiac Muscle
  ↑ Rate and Force of Contraction
    ↑ Cardiac Output
    ↑ in Local Metabolites
    Dilates Coronaries

Relaxes Smooth Muscle in Wall of Bronchioles → better supply of air to alveoli.

Stimulates Respiration.

Inhibits Movements of Digestive Tract.
Contracts Sphincters of Gut.

Inhibits Wall of Urinary Bladder.
Contracts Ureters and Sphincter of
      Urinary Bladder.

Mobilizes Muscle and Liver Glycogen
    → increase in Blood Sugar,
& mobilizes depot fat → ↑ free fatty acid.

Stimulates Metabolism → ↑ B.M.R.

Exerts favourable effect on contracting Skeletal Muscle → Fatigues less readily.

Increases Coagulability of Blood.

Most of these effects can also be produced by stimulating Sympathetic Nerve Fibres.

The Adrenal Medullae are not essential to life — but without them the body is less able to face emergencies and conditions of stress.

DEVELOPMENT of PITUITARY

The Pituitary Gland consists of ANTERIOR and POSTERIOR parts which differ in ORIGIN, STRUCTURE and FUNCTION.

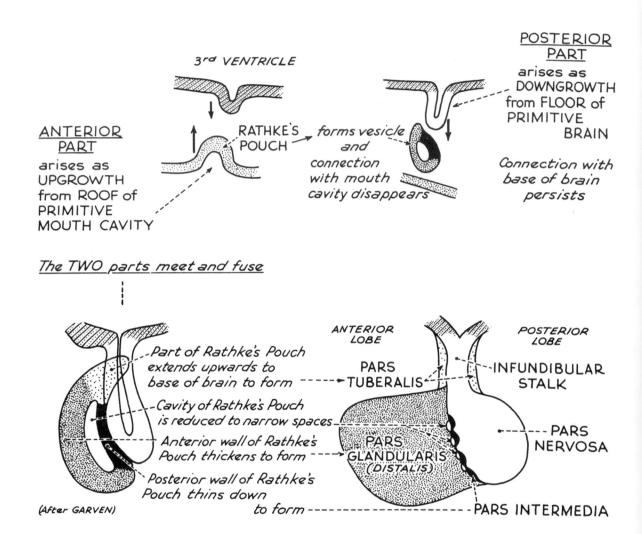

3rd VENTRICLE

POSTERIOR PART arises as DOWNGROWTH from FLOOR of PRIMITIVE BRAIN

Connection with base of brain persists

ANTERIOR PART arises as UPGROWTH from ROOF of PRIMITIVE MOUTH CAVITY

RATHKE'S POUCH → *forms vesicle and connection with mouth cavity disappears*

The TWO parts meet and fuse

Part of Rathke's Pouch extends upwards to base of brain to form ---→ PARS TUBERALIS

Cavity of Rathke's Pouch is reduced to narrow spaces

Anterior wall of Rathke's Pouch thickens to form ---→ PARS GLANDULARIS (DISTALIS)

Posterior wall of Rathke's Pouch thins down to form --- PARS INTERMEDIA

(After GARVEN)

ANTERIOR LOBE

POSTERIOR LOBE

INFUNDIBULAR STALK

PARS NERVOSA

The ADULT PITUITARY (HYPOPHYSIS) is a small oval gland which lies in the SELLA TURCICA — a small cavity in the bone at the base of the skull.

ANTERIOR PITUITARY

This is the MASTER GLAND of the ENDOCRINE SYSTEM.
It regulates the activity of the other Endocrine
Glands, including the GONADS, and influences
ALL METABOLIC PROCESSES
including GROWTH.

CORTEX

STRESS
NERVOUS REFLEXES

HYPOTHALAMUS

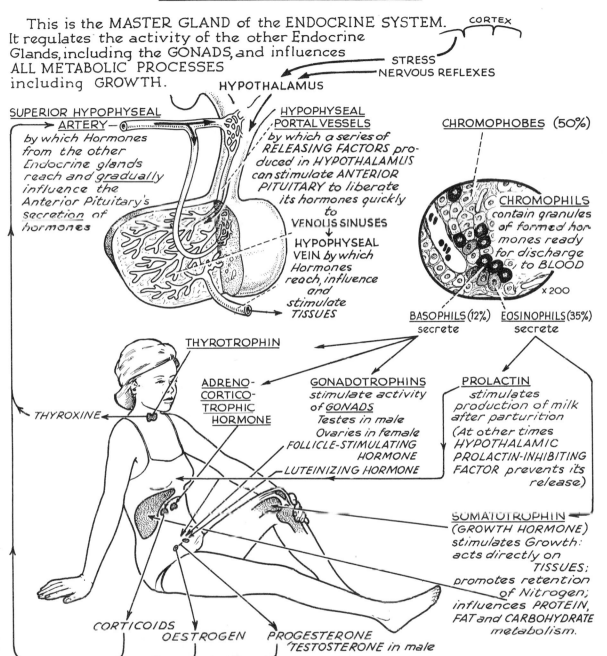

SUPERIOR HYPOPHYSEAL
ARTERY — by which Hormones
from the other
Endocrine glands
reach and gradually
influence the
Anterior Pituitary's
secretion of
hormones

HYPOPHYSEAL
PORTAL VESSELS
by which a series of
RELEASING FACTORS pro-
duced in HYPOTHALAMUS
can stimulate ANTERIOR
PITUITARY to liberate
its hormones quickly
to
VENOUS SINUSES

HYPOPHYSEAL
VEIN by which
Hormones
reach, influence
and
stimulate
TISSUES

CHROMOPHOBES (50%)

CHROMOPHILS
contain granules
of formed hor-
mones ready
for discharge
to BLOOD

x 200

BASOPHILS (12%)
secrete

EOSINOPHILS (35%)
secrete

THYROTROPHIN

THYROXINE

ADRENO-
CORTICO-
TROPHIC
HORMONE

GONADOTROPHINS
stimulate activity
of GONADS
Testes in male
Ovaries in female
FOLLICLE-STIMULATING
HORMONE
LUTEINIZING HORMONE

PROLACTIN
stimulates
production of milk
after parturition
(At other times
HYPOTHALAMIC
PROLACTIN-INHIBITING
FACTOR prevents its
release)

SOMATOTROPHIN
(GROWTH HORMONE)
stimulates Growth;
acts directly on
TISSUES;
promotes retention
of Nitrogen;
influences PROTEIN,
FAT and CARBOHYDRATE
metabolism.

CORTICOIDS
OESTROGEN
PROGESTERONE
TESTOSTERONE in male

UNDERACTIVITY of ANTERIOR PITUITARY

Deficiency or absence
of **EOSINOPHIL** cells

↓

Underproduction of
GROWTH Hormone
(*Somatotrophin*)

↓

<u>LORAIN DWARF</u>

Delayed Skeletal
Growth and
Retarded Sexual
Development
but alert, intelligent,
well proportioned
child.

Destructive disease of part of Anterior
Pituitary(usually with damage to
Posterior Pituitary and/or Hypothalamus)

↓

Underproduction of GROWTH and other
ENDOCRINE-TROPHIC Hormones

↓

FRÖHLICH'S DWARF

Stunting of Growth,
Obesity (*Large
appetite for sugar*);
Arrested Sexual
Development;
Lethargic;
Somnolent;
Mentally
Subnormal.

If Atrophy of
other
Endocrine
glands

↓

Signs of
deficiency
of their
hormones.

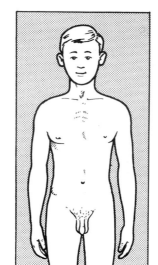
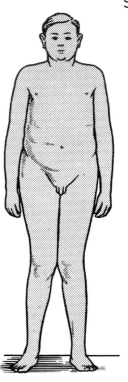

AGE 13

NORMAL CHILD
AGE 13

AGE 13

*GROWTH hormone restores
growth and development
pattern to normal.*

*A similar condition occurs in ADULT
without dwarfing but with suppression
of sex functions and regression of
secondary sex characteristics.
GROWTH and GONADOTROPHIC hor-
mones aid in restoring patient to normal.*

OVERACTIVITY of PITUITARY EOSINOPHIL CELLS

Functional overactivity (or tumour) chiefly of the EOSINOPHIL cells of the Anterior Pituitary leads to ⟶ **GIANTISM** in the CHILD: **ACROMEGALY** in the ADULT.

Overproduction of GROWTH Hormone

General Circulation

Increases NITROGEN retention. Influences Protein, Carbohydrate and Fat metabolism of ALL CELLS of the body.

Overgrowth of all Body Tissues

Onset before bony epiphyses have closed at puberty

Onset after puberty

Bones thicken, especially of FACE, JAW, NOSE, HANDS and FEET ⟶

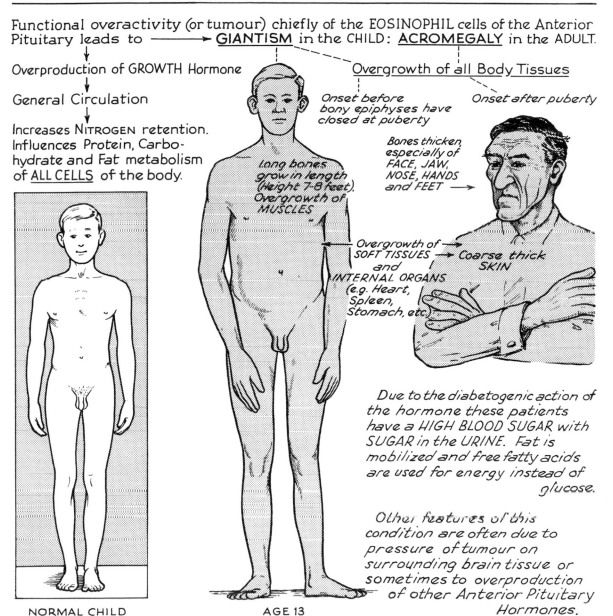

Long bones grow in length (Height 7-8 feet). Overgrowth of MUSCLES

← Overgrowth of ⟶ Coarse thick SOFT TISSUES ⟶ SKIN and INTERNAL ORGANS (e.g. Heart, Spleen, Stomach, etc.)

Due to the diabetogenic action of the hormone these patients have a HIGH BLOOD SUGAR with SUGAR in the URINE. Fat is mobilized and free fatty acids are used for energy instead of glucose.

Other features of this condition are often due to pressure of tumour on surrounding brain tissue or sometimes to overproduction of other Anterior Pituitary Hormones.

NORMAL CHILD AGE 13

AGE 13

Destruction of the overactive tissue – usually by RADIUM therapy – prevents progression of the condition.

OVERACTIVITY of PITUITARY BASOPHIL CELLS

Overactivity *(often due to Tumour)* of the Basophil cells of the Anterior Pituitary gives

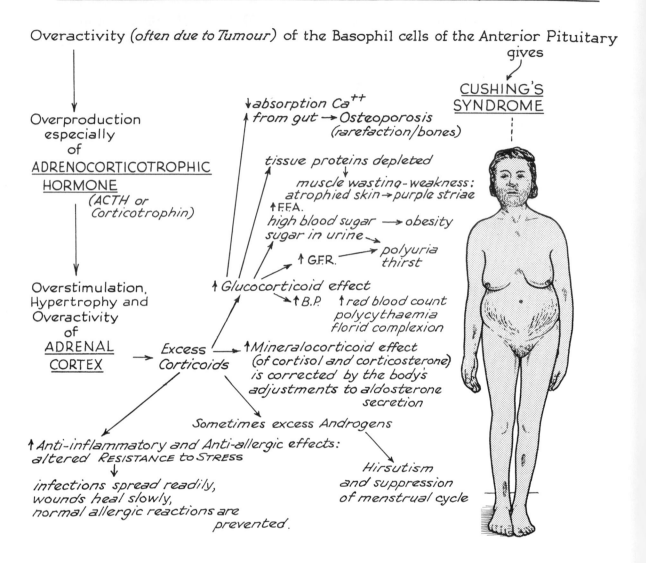

CUSHING'S SYNDROME

Overproduction especially of ADRENOCORTICOTROPHIC HORMONE *(ACTH or Corticotrophin)*

↓absorption Ca^{++}
↑ from gut → Osteoporosis *(rarefaction/bones)*

tissue proteins depleted
muscle wasting-weakness: atrophied skin → purple striae
↑F.F.A.
high blood sugar → obesity
sugar in urine
↑ G.F.R. → polyuria thirst

Overstimulation, Hypertrophy and Overactivity of ADRENAL CORTEX

→ Excess Corticoids

↑Glucocorticoid effect
↗↑B.P.   ↑red blood count polycythaemia florid complexion

↑Mineralocorticoid effect *(of cortisol and corticosterone) is corrected by the body's adjustments to aldosterone secretion*

Sometimes excess Androgens

↑Anti-inflammatory and Anti-allergic effects: altered RESISTANCE to STRESS
↓
infections spread readily, wounds heal slowly, normal allergic reactions are prevented.

Hirsutism and suppression of menstrual cycle

This condition is usually indistinguishable clinically from that seen in primary overactivity or tumour of the Adrenal Cortex itself. The syndrome is here shown in the adult woman.

Overproduction of THYROTROPHIN ⟶ *Overactivity* of THYROID *gland.*

PANHYPOPITUITARISM

Complete Atrophy (or insufficiency) of all secreting cells of Anterior Pituitary
in Adult — SIMMOND'S DISEASE

FAILURE to
PRODUCE
ANY HORMONES ——→ _Features usually
associated with
very OLD AGE_

APPEARANCE of PREMATURE SENILITY

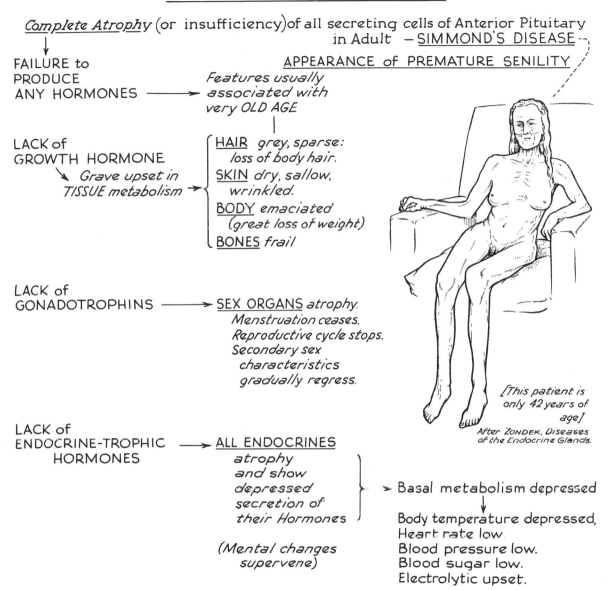

LACK of
GROWTH HORMONE
↘ _Grave upset in
TISSUE metabolism_ ➔

{

HAIR _grey, sparse:
loss of body hair._

SKIN _dry, sallow,
wrinkled._

BODY _emaciated
(great loss of weight)_

BONES _frail_

LACK of
GONADOTROPHINS ——→ SEX ORGANS _atrophy.
Menstruation ceases.
Reproductive cycle stops.
Secondary sex
characteristics
gradually regress._

[This patient is
only 42 years of
age]
After ZONDEK, _Diseases
of the Endocrine Glands._

LACK of
ENDOCRINE-TROPHIC ——→ ALL ENDOCRINES
HORMONES
_atrophy
and show
depressed
secretion of
their Hormones_

}

➤ Basal metabolism depressed
↓
Body temperature depressed.
Heart rate low.
Blood pressure low.
Blood sugar low.
Electrolytic upset.

_(Mental changes
supervene)_

_Anterior Pituitary hormones may relieve the condition but rarely
succeed in completely restoring the patient to normal._

167

POSTERIOR PITUITARY

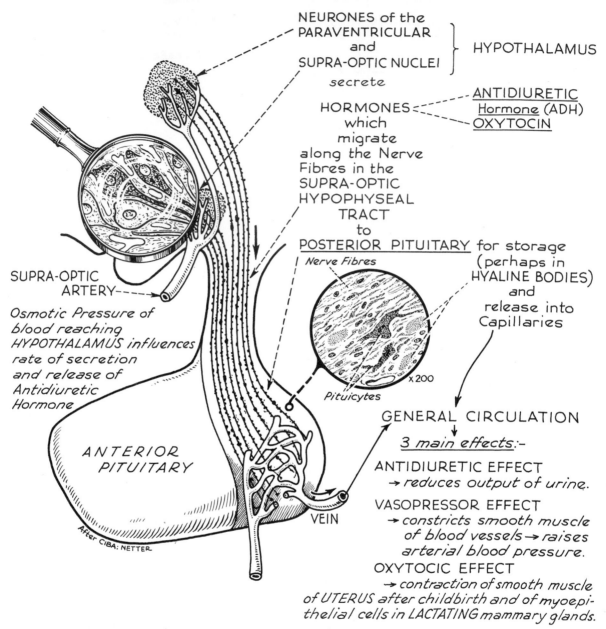

NEURONES of the PARAVENTRICULAR and SUPRA-OPTIC NUCLEI

HYPOTHALAMUS

secrete

HORMONES which migrate along the Nerve Fibres in the SUPRA-OPTIC HYPOPHYSEAL TRACT to POSTERIOR PITUITARY for storage (perhaps in HYALINE BODIES) and release into Capillaries

ANTIDIURETIC Hormone (ADH)

OXYTOCIN

Nerve Fibres

Pituicytes

x200

SUPRA-OPTIC ARTERY

Osmotic Pressure of blood reaching HYPOTHALAMUS influences rate of secretion and release of Antidiuretic Hormone

ANTERIOR PITUITARY

After CIBA: NETTER

VEIN

GENERAL CIRCULATION

3 main effects:-

ANTIDIURETIC EFFECT
→ reduces output of urine.

VASOPRESSOR EFFECT
→ constricts smooth muscle of blood vessels → raises arterial blood pressure.

OXYTOCIC EFFECT
→ contraction of smooth muscle of UTERUS after childbirth and of myoepithelial cells in LACTATING mammary glands.

There are no obvious secreting cells in Posterior Pituitary such as are seen in ENDOCRINE GLANDS

OXYTOCIN

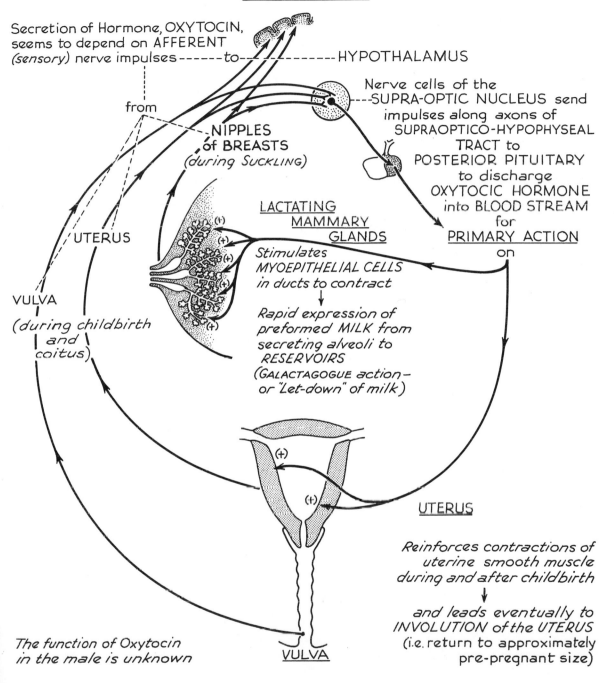

Secretion of Hormone, OXYTOCIN, seems to depend on AFFERENT (sensory) nerve impulses - - - - - - to - - - - - - - - - - - - HYPOTHALAMUS

from

NIPPLES of BREASTS (during Suckling)

Nerve cells of the SUPRA-OPTIC NUCLEUS send impulses along axons of SUPRAOPTICO-HYPOPHYSEAL TRACT to POSTERIOR PITUITARY to discharge OXYTOCIC HORMONE into BLOOD STREAM for PRIMARY ACTION on

UTERUS

VULVA (during childbirth and coitus)

LACTATING MAMMARY GLANDS

Stimulates MYOEPITHELIAL CELLS in ducts to contract

Rapid expression of preformed MILK from secreting alveoli to RESERVOIRS (Galactagogue action — or "Let-down" of milk)

UTERUS

Reinforces contractions of uterine smooth muscle during and after childbirth

and leads eventually to INVOLUTION of the UTERUS (i.e. return to approximately pre-pregnant size)

VULVA

The function of Oxytocin in the male is unknown

ANTIDIURETIC HORMONE

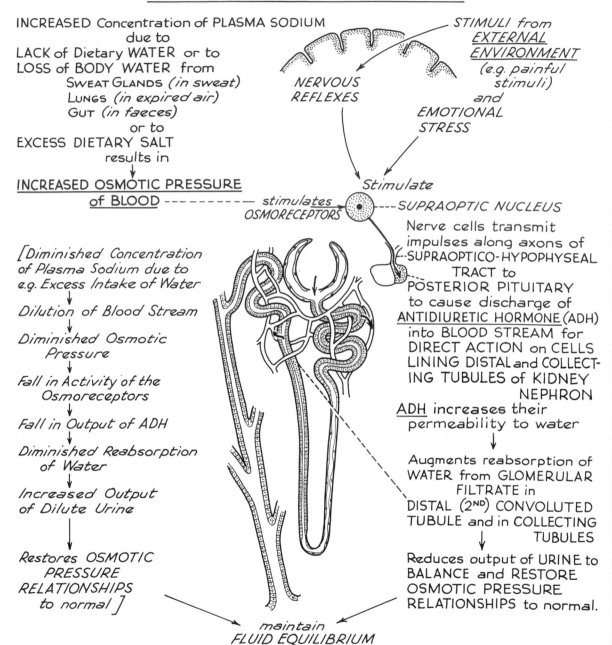

INCREASED Concentration of PLASMA SODIUM
due to
LACK of Dietary WATER or to
LOSS of BODY WATER from
    Sweat Glands (in sweat)
    Lungs (in expired air)
    Gut (in faeces)
        or to
EXCESS DIETARY SALT
results in

INCREASED OSMOTIC PRESSURE
of BLOOD – – – – – – – – stimulates
                        OSMORECEPTORS

STIMULI from
EXTERNAL
ENVIRONMENT
(e.g. painful
stimuli)
and
EMOTIONAL
STRESS

NERVOUS
REFLEXES

Stimulate

–SUPRAOPTIC NUCLEUS

Nerve cells transmit
impulses along axons of
SUPRAOPTICO-HYPOPHYSEAL
TRACT to
POSTERIOR PITUITARY
to cause discharge of
ANTIDIURETIC HORMONE (ADH)
into BLOOD STREAM for
DIRECT ACTION on CELLS
LINING DISTAL and COLLECT-
ING TUBULES of KIDNEY
NEPHRON
ADH increases their
permeability to water

Augments reabsorption of
WATER from GLOMERULAR
FILTRATE in
DISTAL (2ND) CONVOLUTED
TUBULE and in COLLECTING
TUBULES

Reduces output of URINE to
BALANCE and RESTORE
OSMOTIC PRESSURE
RELATIONSHIPS to normal.

[Diminished Concentration
of Plasma Sodium due to
e.g. Excess Intake of Water

Dilution of Blood Stream

Diminished Osmotic
Pressure

Fall in Activity of the
Osmoreceptors

Fall in Output of ADH

Diminished Reabsorption
of Water

Increased Output
of Dilute Urine

Restores OSMOTIC
PRESSURE
RELATIONSHIPS
to normal]

maintain
FLUID EQUILIBRIUM

(This hormone also has vasoconstrictor effects ⟶ increased Blood Pressure)

UNDERACTIVITY of POSTERIOR PITUITARY

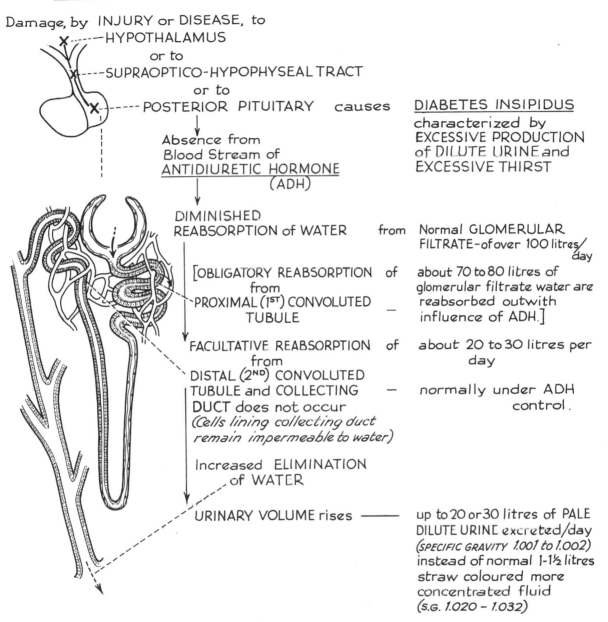

Damage, by INJURY or DISEASE, to

X------HYPOTHALAMUS
            or to
X------SUPRAOPTICO-HYPOPHYSEAL TRACT
            or to
X------POSTERIOR PITUITARY   causes   DIABETES INSIPIDUS
                                        characterized by
            Absence from   EXCESSIVE PRODUCTION
            Blood Stream of   of DILUTE URINE and
            ANTIDIURETIC HORMONE   EXCESSIVE THIRST
                (ADH)

DIMINISHED
REABSORPTION of WATER   from   Normal GLOMERULAR
                                  FILTRATE—of over 100 litres/
                                                            day

[OBLIGATORY REABSORPTION of   about 70 to 80 litres of
            from   glomerular filtrate water are
PROXIMAL (1ST) CONVOLUTED   —   reabsorbed outwith
            TUBULE   influence of ADH.]

FACULTATIVE REABSORPTION of   about 20 to 30 litres per
            from   day
DISTAL (2ND) CONVOLUTED
TUBULE and COLLECTING   —   normally under ADH
DUCT does not occur   control.
(Cells lining collecting duct
remain impermeable to water)

Increased ELIMINATION
    of WATER

URINARY VOLUME rises ———   up to 20 or 30 litres of PALE
                                  DILUTE URINE excreted/day
                                  (SPECIFIC GRAVITY 1.001 to 1.002)
                                  instead of normal 1–1½ litres
                                  straw coloured more
                                  concentrated fluid
                                  (S.G. 1.020 – 1.032)

Small amounts of Posterior Pituitary extract absorbed from under the tongue or given by subcutaneous injection reduce elimination of water to normal.

ALDOSTERONE and ANTIDIURETIC HORMONE (ADH) in the MAINTENANCE of BLOOD VOLUME

A reduction in the total volume of EXTRACELLULAR FLUID (e.g. after haemorrhage or loss of isotonic secretions from the gut in vomiting or diarrhoea) leads to chain of COMPENSATORY mechanisms in which ALDOSTERONE plays an important role.

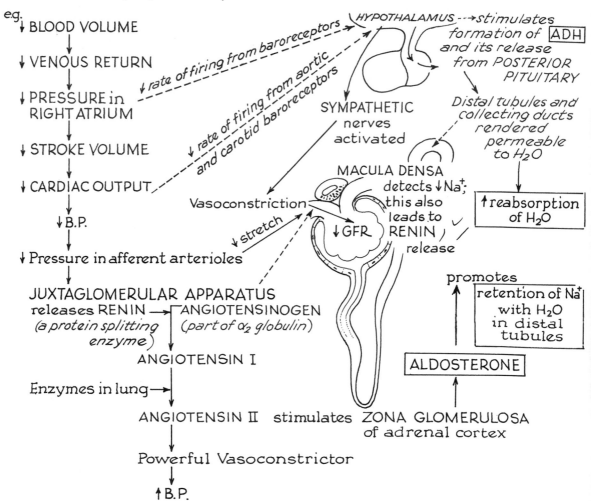

eg.
↓ BLOOD VOLUME

↓ VENOUS RETURN

↓ PRESSURE in RIGHT ATRIUM

↓ STROKE VOLUME

↓ CARDIAC OUTPUT

↓ B.P.

↓ Pressure in afferent arterioles

JUXTAGLOMERULAR APPARATUS
releases RENIN → ANGIOTENSINOGEN
(a protein splitting enzyme)   *(part of α_2 globulin)*

ANGIOTENSIN I

Enzymes in lung →

ANGIOTENSIN II   stimulates   ZONA GLOMERULOSA of adrenal cortex

Powerful Vasoconstrictor

↑ B.P.

↓ rate of firing from baroreceptors

↓ rate of firing from aortic and carotid baroreceptors

HYPOTHALAMUS ---→ stimulates formation of ADH and its release from POSTERIOR PITUITARY

SYMPATHETIC nerves activated

Vasoconstriction

↓ stretch

MACULA DENSA detects ↓ Na^+; this also leads to RENIN release

↓ GFR

Distal tubules and collecting ducts rendered permeable to H_2O

↑ reabsorption of H_2O

promotes retention of Na^+ with H_2O in distal tubules

ALDOSTERONE

These measures serve to maintain BLOOD VOLUME till the long term replacement of the lost r.b.c., plasma poteins and electrolytes can be achieved.

172

PANCREAS: ISLETS of LANGERHANS

ISLETS of LANGERHANS make up 1-2% of Pancreatic Tissue.

×150

α cells secrete GLUCAGON —— *ACTIONS*
(25%)   Promotes GLYCOGEN breakdown

↓ BLOOD SUGAR
(Together with GROWTH HORMONE,
GLUCOCORTICOIDS and ADRENALINE,
co-operates in mobilizing energy
stores during the FASTING STATE);
protects against HYPOGLYCAEMIA.

CONTROL
↓ BLOOD SUGAR   ↑ BLOOD
(fasting =   SUGAR
↓ 80-90mg%)
↑ GLUCAGON   ↓ GLUCAGON

Breakdown of
GLYCOGEN

β cells (50-75%) secrete INSULIN → Enhances uptake and utilization
of glucose by muscle and
adipose tissue.

CONTROL is by "negative feedback" with BLOOD GLUCOSE

↑ BLOOD GLUCOSE   ↓ BLOOD   *ACTIONS*
after a meal   GLUCOSE   Promotes
                                      GLYCOGENESIS in liver and muscle ⎫ ↓ BLOOD
                                      LIPOGENESIS - glucose converted to │ GLUCOSE
                                                  fat and stored   │
↑ INSULIN   ↓ INSULIN   Uptake and utilization of GLUCOSE ⎬
                                         by muscle and adipose tissue │ ↑ FAT and
                                      Inhibits   │ GLYCOGEN
                                      GLYCOGENOLYSIS   │ STORES
                                      LIPOLYSIS - utilization of fat for energy ⎭

ATROPHY of ISLETS → ABSENCE of INSULIN → inhibits GLUCOSE UTILIZATION
DIABETES MELLITUS ←   ↘ promotes FAT UTILIZATION
↓↓ rate of glucose transport across cell
                            membranes   ↑↑ BLOOD SUGAR *(may reach*
↓↓ UTILIZATION of GLUCOSE by MUSCLE   *180mg/100ml)*

                                             Renal threshold exceeded

↑↑ MOBILIZATION and   ⎫   GLUCOSE in URINE *(lost in solution*
UTILIZATION of DEPOT FAT ⎬ → ↑↑ Free fatty acids   *with H₂0)*
instead of glucose   ⎭   and ketone bodies
                             in blood ————→   POLYURIA and POLYPHAGIA
                                            → THIRST ++ → POLYDIPSIA
FAT STORES depleted   KETONE BODIES in URINE
FAT DEPOSITS in BLOOD VESSELS
   (atherosclerosis)
↑↑ GLUCONEOGENESIS → DEPLETION of BODY PROTEIN → MUSCLE WASTING and
                                                 LOSS of WEIGHT
                                                 FATIGUE readily

If untreated → progressive → drowsiness ——→ coma ——→ death

Excess Insulin (HYPERINSULINISM) → low blood sugar → irritability; sweating; hunger.
If untreated → reduction of metabolism of nervous tissues → giddiness → coma → death.

173

THYMUS

The Thymus is an irregularly-shaped organ lying behind the breast bone. It is relatively large in the child and reaches its maximum size at puberty. It closely resembles a Lymph Node.

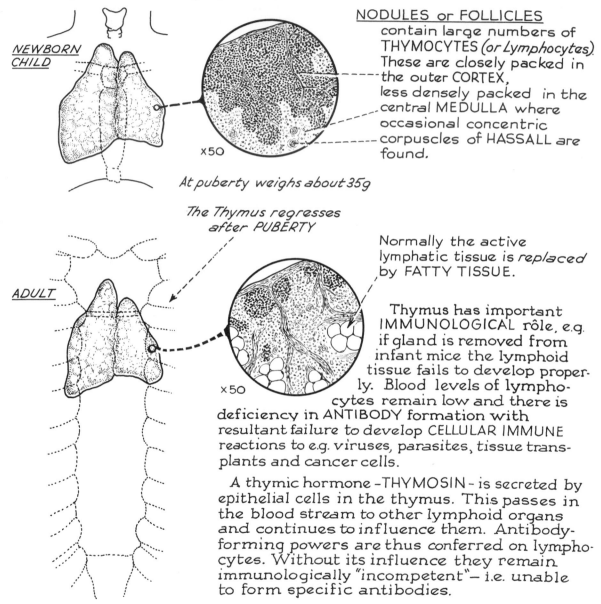

NEWBORN CHILD

x50

At puberty weighs about 35g

The Thymus regresses after PUBERTY

ADULT

x50

NODULES or FOLLICLES
contain large numbers of THYMOCYTES *(or Lymphocytes)* These are closely packed in the outer CORTEX, less densely packed in the central MEDULLA where occasional concentric corpuscles of HASSALL are found.

Normally the active lymphatic tissue is *replaced* by FATTY TISSUE.

Thymus has important IMMUNOLOGICAL rôle, e.g. if gland is removed from infant mice the lymphoid tissue fails to develop properly. Blood levels of lymphocytes remain low and there is deficiency in ANTIBODY formation with resultant failure to develop CELLULAR IMMUNE reactions to e.g. viruses, parasites, tissue transplants and cancer cells.

A thymic hormone –THYMOSIN– is secreted by epithelial cells in the thymus. This passes in the blood stream to other lymphoid organs and continues to influence them. Antibody-forming powers are thus conferred on lymphocytes. Without its influence they remain immunologically "incompetent"– i.e. unable to form specific antibodies.

CHAPTER 8

REPRODUCTIVE SYSTEM

MALE REPRODUCTIVE SYSTEM

<u>PRIMARY SEX ORGANS</u> *produce* the *MALE GERM CELLS* — <u>SPERMATOZOA</u>
  TESTES (Two) *and the MALE SEX HORMONE* —
                        <u>TESTOSTERONE</u>

Testosterone is responsible for
development at Puberty of :-
  <u>SECONDARY SEX ORGANS</u>
        EPIDIDYMIS (Two) ———
        VAS DEFERENS (Two) ——— } *transfer Spermatozoa from the Testes.*
        SEMINAL VESICLES (Two) — } *secrete fluid medium for transport of Spermatozoa.*
        PROSTATE GLAND ———
        PENIS ——————— *transfers Spermatozoa from male to female.*

  and
appearance of
<u>SECONDARY SEX CHARACTERISTICS</u>
Laryngeal changes → Deep voice.
Pubic, Axillary and Facial hair.
Characteristic Male shape of
                    body.

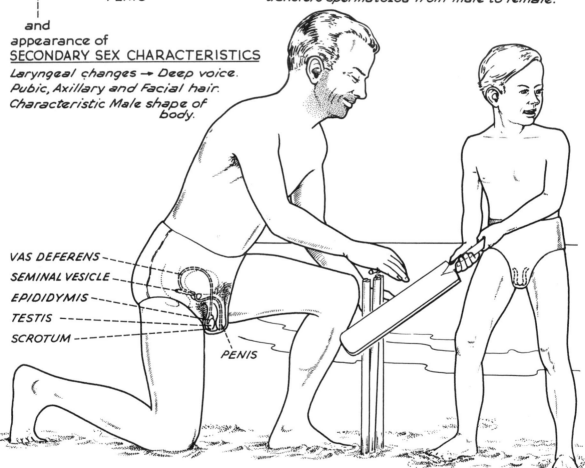

VAS DEFERENS
SEMINAL VESICLE
EPIDIDYMIS
TESTIS
SCROTUM
                    PENIS

In the male the process of spermatogenesis starts just after
                    puberty and is normally continuous until old age.

176

TESTIS

There are TWO TESTES.
These produce the Male GERM CELLS:—

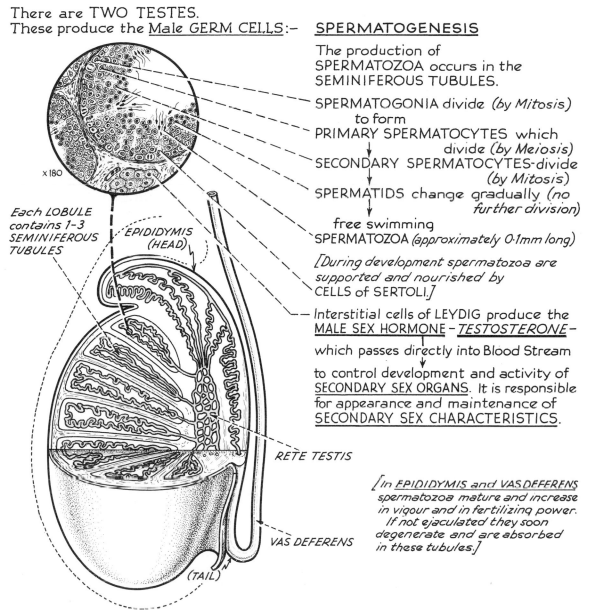

x180

Each LOBULE
contains 1–3
SEMINIFEROUS
TUBULES

EPIDIDYMIS
(HEAD)

RETE TESTIS

VAS DEFERENS

(TAIL)

SPERMATOGENESIS

The production of
SPERMATOZOA occurs in the
SEMINIFEROUS TUBULES.

SPERMATOGONIA divide *(by Mitosis)*
to form
PRIMARY SPERMATOCYTES which
divide *(by Meiosis)*
SECONDARY SPERMATOCYTES–divide
(by Mitosis)
SPERMATIDS change gradually *(no
further division)*
free swimming
SPERMATOZOA *(approximately 0·1mm long)*

*[During development spermatozoa are
supported and nourished by
CELLS of SERTOLI.]*

Interstitial cells of LEYDIG produce the
MALE SEX HORMONE – *TESTOSTERONE* –

which passes directly into Blood Stream

to control development and activity of
SECONDARY SEX ORGANS. It is responsible
for appearance and maintenance of
SECONDARY SEX CHARACTERISTICS.

*[In EPIDIDYMIS and VAS DEFERENS
spermatozoa mature and increase
in vigour and in fertilizing power.
If not ejaculated they soon
degenerate and are absorbed
in these tubules.]*

*Events occurring in the testes are under CONTROL of HORMONES chiefly
those of ANTERIOR PITUITARY and the HYPOTHALAMUS.*

MALE SECONDARY SEX ORGANS

These are the organs adapted for TRANSFER of live SPERMATOZOA from male to female.

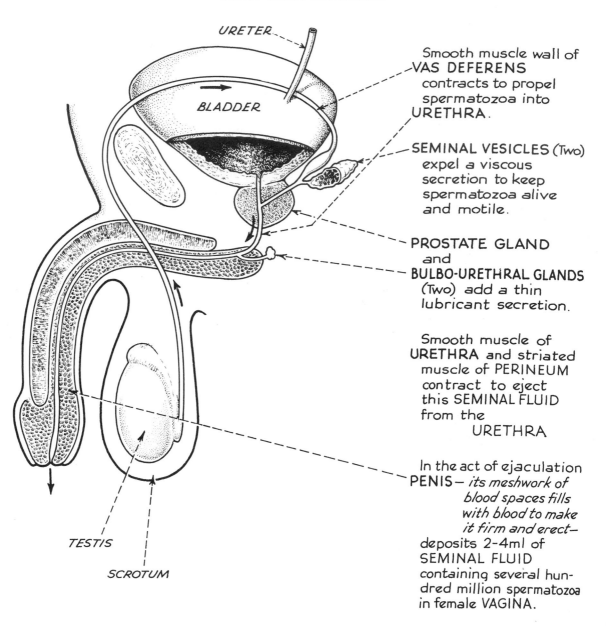

URETER

BLADDER

Smooth muscle wall of **VAS DEFERENS** contracts to propel spermatozoa into **URETHRA**.

SEMINAL VESICLES (Two) expel a viscous secretion to keep spermatozoa alive and motile.

PROSTATE GLAND and **BULBO-URETHRAL GLANDS** (Two) add a thin lubricant secretion.

Smooth muscle of **URETHRA** and striated muscle of PERINEUM contract to eject this SEMINAL FLUID from the
URETHRA

In the act of ejaculation **PENIS** — *its meshwork of blood spaces fills with blood to make it firm and erect* — deposits 2-4ml of SEMINAL FLUID containing several hundred million spermatozoa in female VAGINA.

TESTIS

SCROTUM

CONTROL OF EVENTS IN THE TESTIS

Between the ages of 13 and 16 years the hypothalamus begins to secrete GONADOTROPHIN-RELEASING HORMONE (GnRH).

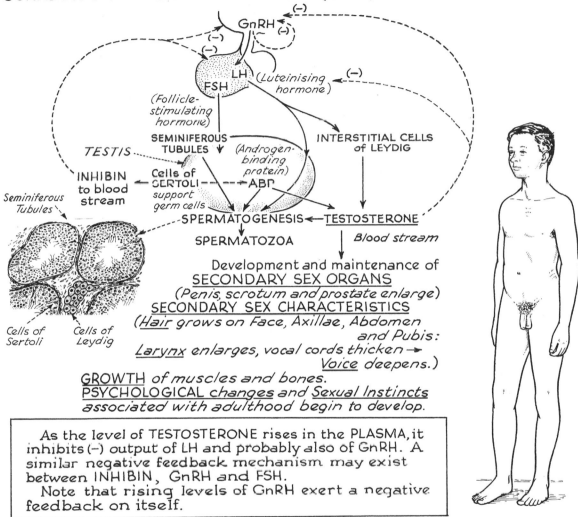

Development and maintenance of
SECONDARY SEX ORGANS
(Penis, scrotum and prostate enlarge)
SECONDARY SEX CHARACTERISTICS
(Hair grows on Face, Axillae, Abdomen
and Pubis:
Larynx enlarges, vocal cords thicken →
Voice deepens.)

GROWTH *of muscles and bones.*
PSYCHOLOGICAL *changes and Sexual Instincts*
associated with adulthood begin to develop.

As the level of TESTOSTERONE rises in the PLASMA, it inhibits (–) output of LH and probably also of GnRH. A similar negative feedback mechanism may exist between INHIBIN, GnRH and FSH.

Note that rising levels of GnRH exert a negative feedback on itself.

If Atrophy of Testes occurs → *at time of normal puberty* → Sex organs remain small.
Secondary sex characteristics fail to develop.
after puberty → Spermatogenesis stops → Sterility.
Testosterone production falls → Atrophy of secondary sex organs.

Injections of Testosterone in cases of delayed puberty → changes associated with puberty.

FEMALE REPRODUCTIVE SYSTEM

<u>PRIMARY SEX ORGANS</u>
OVARIES (Two)

produce the FEMALE GERM CELLS — <u>OVA</u>
and the FEMALE SEX HORMONES —
<u>OESTROGEN</u> and <u>PROGESTERONE</u>

Oestrogen and Progesterone are responsible
for development at Puberty of:-
<u>SECONDARY SEX ORGANS</u>

FALLOPIAN TUBES (Two) — for the transfer of the Ova from Ovaries.
VAGINA ——————— for the reception of the Male Germ Cells.
UTERUS ——————— for the nutrition and development of the
fertilized Egg Cell → developing embryo.
MAMMARY GLANDS (Two) - for the nutrition of the New Individual after birth.

and

appearance and
maintenance of
<u>SECONDARY SEX
CHARACTERISTICS</u>

Development of
Breasts

Axillary and Pubic
hair

Typical Feminine
Proportions of body

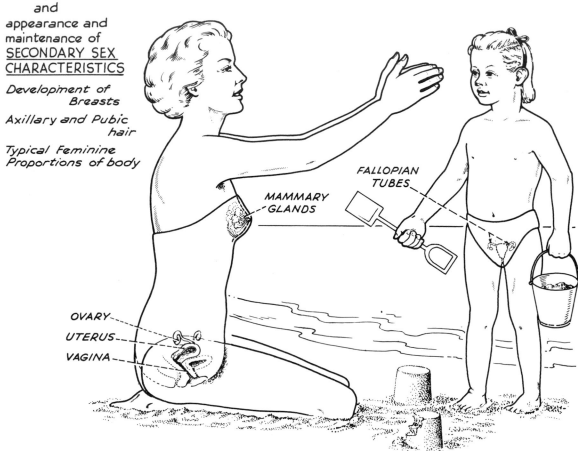

*In the female the cyclical production of ova starts just after puberty and
continues (unless interrupted by pregnancy or disease) until the menopause.*

180

ADULT PELVIC SEX ORGANS
in ORDINARY FEMALE CYCLE

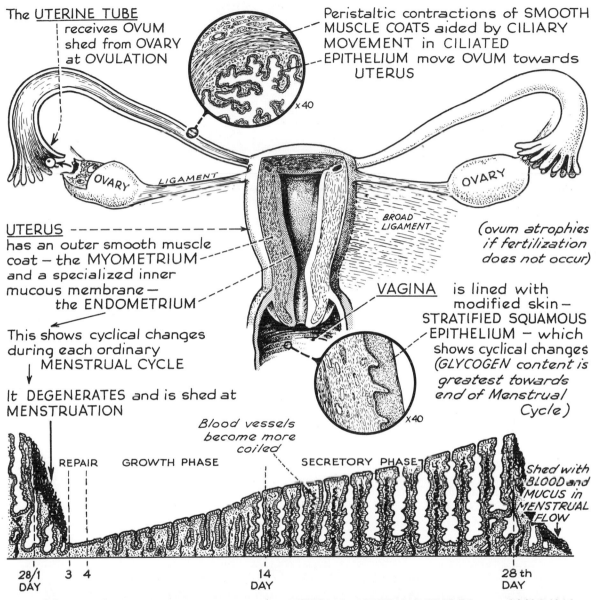

The UTERINE TUBE receives OVUM shed from OVARY at OVULATION

Peristaltic contractions of SMOOTH MUSCLE COATS aided by CILIARY MOVEMENT in CILIATED EPITHELIUM move OVUM towards UTERUS

x 40

OVARY

LIGAMENT

BROAD LIGAMENT

OVARY

(ovum atrophies if fertilization does not occur)

UTERUS — — — — —
has an outer smooth muscle coat — the MYOMETRIUM and a specialized inner mucous membrane — the ENDOMETRIUM

VAGINA is lined with modified skin — STRATIFIED SQUAMOUS EPITHELIUM — which shows cyclical changes (GLYCOGEN content is greatest towards end of Menstrual Cycle)

This shows cyclical changes during each ordinary MENSTRUAL CYCLE

It DEGENERATES and is shed at MENSTRUATION

Blood vessels become more coiled

x 40

REPAIR   GROWTH PHASE   SECRETORY PHASE

Shed with BLOOD and MUCUS in MENSTRUAL FLOW

28/1 DAY   3 4   14 DAY   28 th DAY

Rhythmical changes occur in UTERUS, UTERINE TUBES and VAGINA under action of OVARIAN Hormones.

OVARY in ORDINARY ADULT CYCLE

There are TWO OVARIES. These produce the Female GERM CELLS.
The production of OVA is a CYCLICAL PROCESS — OOGENESIS

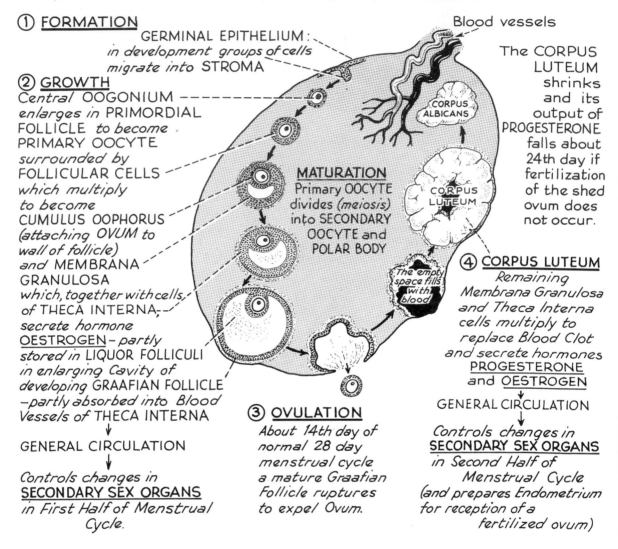

① FORMATION
GERMINAL EPITHELIUM :
*in development groups of cells
migrate into STROMA*

② GROWTH
Central OOGONIUM
enlarges in PRIMORDIAL
FOLLICLE *to become*
PRIMARY OOCYTE
surrounded by
FOLLICULAR CELLS
*which multiply
to become*
CUMULUS OOPHORUS
*(attaching OVUM to
wall of follicle)
and* MEMBRANA
GRANULOSA
*which, together with cells
of* THECA INTERNA
secrete hormone
OESTROGEN *— partly
stored in* LIQUOR FOLLICULI
*in enlarging Cavity of
developing* GRAAFIAN FOLLICLE
*—partly absorbed into Blood
Vessels of* THECA INTERNA
↓
GENERAL CIRCULATION
↓
Controls changes in
SECONDARY SEX ORGANS
*in First Half of Menstrual
Cycle.*

Blood vessels

MATURATION
Primary OOCYTE
*divides (meiosis)
into* SECONDARY
OOCYTE *and*
POLAR BODY

CORPUS
ALBICANS

CORPUS
LUTEUM

*The empty
space fills
with
blood*

The CORPUS
LUTEUM
shrinks
and its
output of
PROGESTERONE
falls about
24th day if
fertilization
of the shed
ovum does
not occur.

④ CORPUS LUTEUM
*Remaining
Membrana Granulosa
and Theca Interna
cells multiply to
replace Blood Clot
and secrete hormones*
PROGESTERONE
and OESTROGEN
↓
GENERAL CIRCULATION
↓
Controls changes in
SECONDARY SEX ORGANS
*in Second Half of
Menstrual Cycle
(and prepares Endometrium
for reception of a
fertilized ovum)*

③ OVULATION
*About 14th day of
normal 28 day
menstrual cycle
a mature Graafian
Follicle ruptures
to expel Ovum.*

*For simplicity the development of only one Graafian Follicle is shown here.
Several grow in each cycle but in the Human subject usually only one Follicle
ruptures. The others atrophy: i.e. ONE MATURE OVUM is shed each month.
Events in the OVARY are under control of ANTERIOR PITUITARY HORMONES
and HYPOTHALAMIC RELEASING FACTORS.*

OVARY in PREGNANCY

When pregnancy occurs the ordinary ovarian cycle is suspended.

After the first 14 days the developing placenta secretes a LUTEINIZING HORMONE (chorionic gonadotrophin).

Under its influence the CORPUS LUTEUM continues to grow until it may come to occupy 30% to 50% of the total volume of the OVARY.

The large amount of PROGESTERONE

helps to maintain the PREGNANCY in its early stages and is essential for development of the PLACENTA – the special structure through which the child receives its nourishment from the mother.
It also produces PROGESTERONE which gradually takes over from the

CORPUS LUTEUM ——

reaches its peak at about 6 weeks after conception

falls off about 2nd. month

ceases to be active about 4th. month

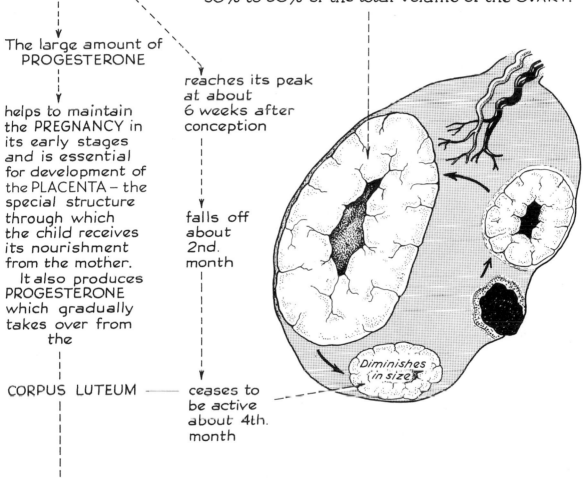

Diminishes in size

PLACENTAL PROGESTERONE —— takes over to maintain the PREGNANCY, and to help to prepare the mammary glands for Lactation.

183

CONTROL OF EVENTS IN THE OVARY

Between the ages of 10 and 14 years the HYPOTHALAMUS begins to secrete GONADOTROPHIN-RELEASING HORMONE (Gn.RH). The girl enters puberty. Thereafter the cycle is controlled thus:-

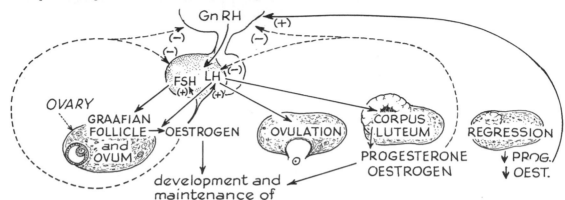

development and maintenance of
SECONDARY SEX ORGANS
(Fallopian tubes lengthen: Uterine muscle enlarges, its lining proliferates and undergoes cyclical changes leading to onset of MENSTRUATION: Vaginal epithelium thickens.)
SECONDARY SEX CHARACTERISTICS
(Hair grows on Axillae and Pubis; Mammary Glands enlarge.)
GROWTH: *Bones grow; Height increases; Pelvis becomes broader; Fat is deposited on shoulders, hips and thighs to give typical feminine contours.*
PSYCHOLOGICAL and PERSONALITY *changes occur.*
i.e. PUBERTY changes GIRL CHILD into WOMAN able to bear children.

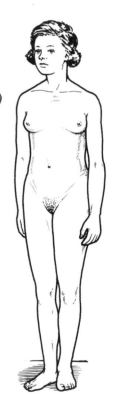

Note: As the level of OESTROGEN rises in the early stages of the cycle it inhibits (–) output of GnRH, FSH and LH.

About the 12th. or 13th. day the high level of oestrogen causes positive feedback (+). A sudden surge of LH (and of FSH) leads to OVULATION and the formation of the CORPUS LUTEUM. As the level of PROGESTERONE rises (along with OESTROGEN) it inhibits (–) GnRH, LH and FSH.

A few days before menstruation the CORPUS LUTEUM involutes. As the levels of PROGESTERONE and OESTROGEN fall, GnRH is freed from inhibition. Levels of FSH and LH begin to rise. The cycle starts again.

OVARIAN HORMONES

The Ovarian Cycle is repeated monthly from PUBERTY to the MENOPAUSE unless interrupted by PREGNANCY or DISEASE.

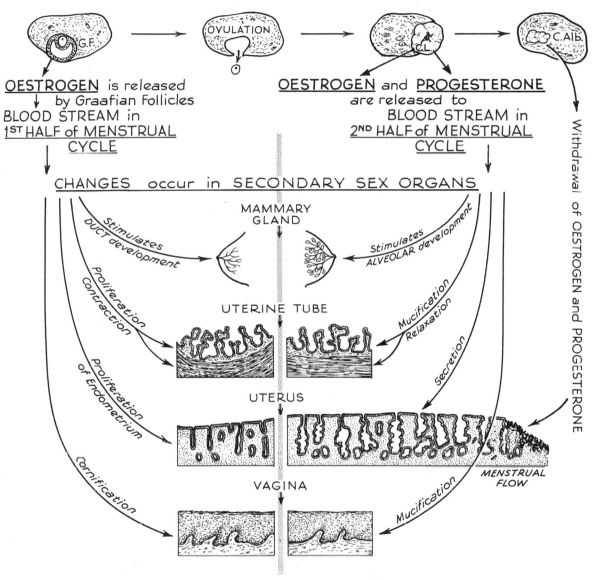

OVARIAN HORMONES are therefore directly responsible for regular cycle of events in SECONDARY SEX ORGANS.

UTERUS and UTERINE TUBES

The UTERUS (or Womb) is the organ which bears the developing child till birth.

In CHILDHOOD -

— it is a small undeveloped organ
situated deep in the pelvis.

OUTER WALLS of
   Smooth muscle —
     the MYOMETRIUM

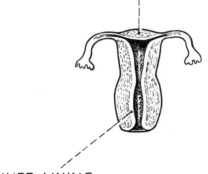

INNER LINING or
   Mucous membrane —
    the ENDOMETRIUM

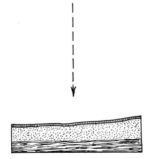

At PUBERTY

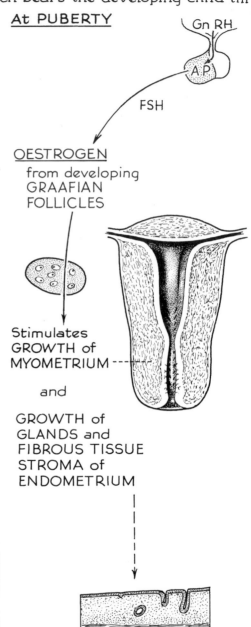

Gn RH

A.P.

FSH

OESTROGEN
  from developing
  GRAAFIAN
  FOLLICLES

Stimulates
GROWTH of
MYOMETRIUM

and

GROWTH of
GLANDS and
FIBROUS TISSUE
STROMA of
ENDOMETRIUM

UTERUS and UTERINE TUBES

In MATURITY
[From Puberty to Menopause (unless interrupted by pregnancy or disease) — when ovarian cycle is fully established]

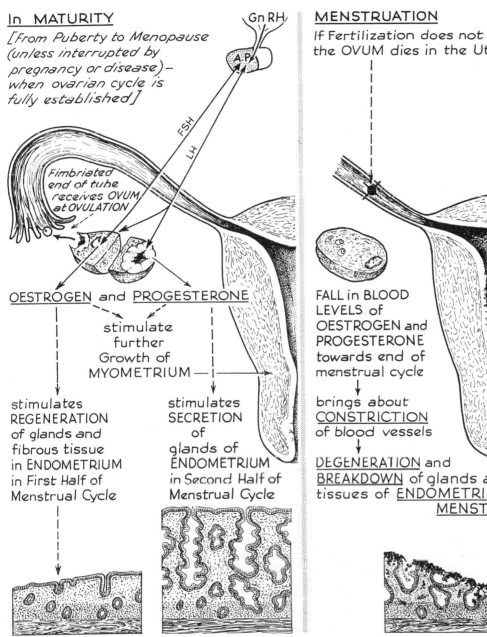

Fimbriated end of tube receives OVUM at OVULATION

Gn RH

A.P.

FSH

LH

OESTROGEN and PROGESTERONE

→ stimulate further Growth of MYOMETRIUM —

stimulates REGENERATION of glands and fibrous tissue in ENDOMETRIUM in First Half of Menstrual Cycle

stimulates SECRETION of glands of ENDOMETRIUM in Second Half of Menstrual Cycle

MENSTRUATION
If Fertilization does not take place the OVUM dies in the Uterine tube

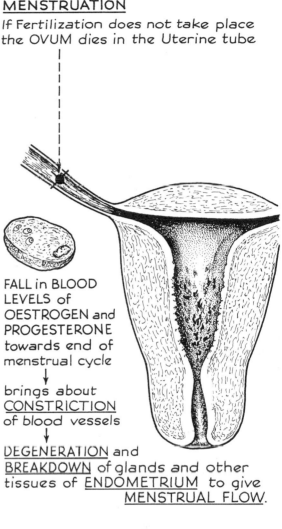

FALL in BLOOD LEVELS of OESTROGEN and PROGESTERONE towards end of menstrual cycle

↓

brings about CONSTRICTION of blood vessels

↓

DEGENERATION and BREAKDOWN of glands and other tissues of ENDOMETRIUM to give MENSTRUAL FLOW.

UTERINE TUBES
in cycle ending in PREGNANCY

The fimbriated end of the UTERINE TUBE receives the OVUM at OVULATION. PERISTALTIC contractions of the muscular tube aided by ciliary movements of its lining cells transfer the OVUM towards the UTERUS. The uterine tube also transmits SPERMATOZOA towards the OVA.

FERTILIZATION
—or fusion of OVUM and SPERM — occurs in outer third of uterine tube

MATURATION
of OVUM now occurs *(the Polar bodies finally degenerate)*

CLEAVAGE
After fertilization in the uterine tube the fertilized Ovum or ZYGOTE undergoes several divisions

BLASTULATION

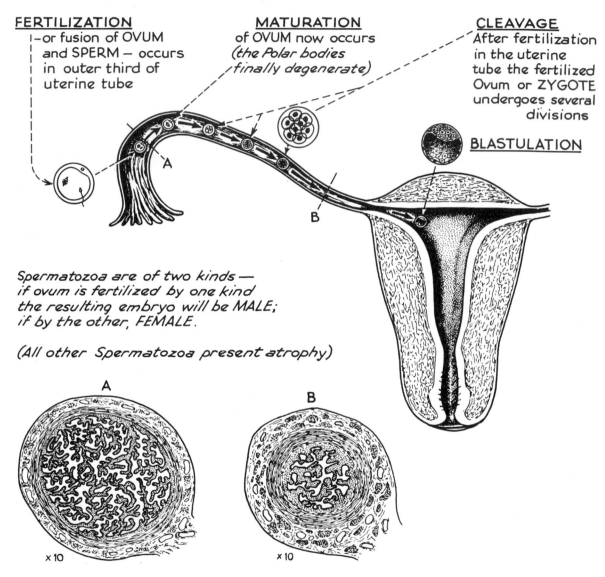

Spermatozoa are of two kinds — if ovum is fertilized by one kind the resulting embryo will be MALE; if by the other, FEMALE.

(All other Spermatozoa present atrophy)

A ×10

B ×10

UTERUS

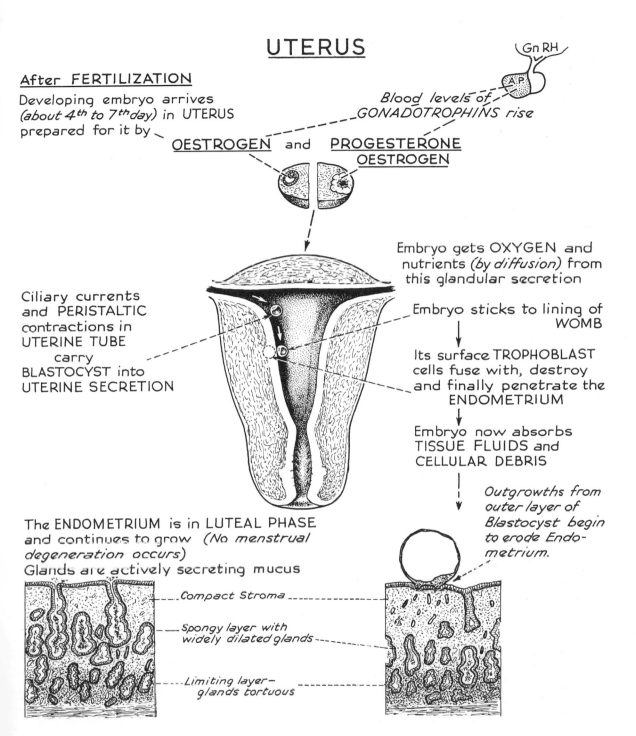

After FERTILIZATION

Developing embryo arrives
(about 4th to 7th day) in UTERUS
prepared for it by

OESTROGEN and **PROGESTERONE**
OESTROGEN

Gn RH
A.P.

Blood levels of
GONADOTROPHINS rise

Embryo gets OXYGEN and
nutrients *(by diffusion)* from
this glandular secretion

Ciliary currents
and PERISTALTIC
contractions in
UTERINE TUBE
carry
BLASTOCYST into
UTERINE SECRETION

Embryo sticks to lining of
WOMB

Its surface TROPHOBLAST
cells fuse with, destroy
and finally penetrate the
ENDOMETRIUM

Embryo now absorbs
TISSUE FLUIDS and
CELLULAR DEBRIS

Outgrowths from
outer layer of
Blastocyst begin
to erode Endo-
metrium.

The ENDOMETRIUM is in LUTEAL PHASE
and continues to grow *(No menstrual*
degeneration occurs)
Glands are actively secreting mucus

Compact Stroma

Spongy layer with
widely dilated glands

Limiting layer-
glands tortuous

189

UTERUS

GnRH

A.P.

IMPLANTATION

Embryo invades
ENDOMETRIUM
*(which is now called
the DECIDUA)*

*Blood levels of
GONADOTROPHINS
continue to rise*

PLACENTATION

PROGESTERONE is necessary
for the development of the
PLACENTA – the special organ
through which the developing
child receives nourishment
from the mother

OESTROGEN and PROGESTERONE

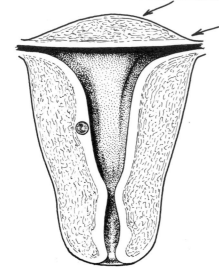

*The HORMONES
of the PLACENTA
help to maintain
the
PREGNANCY*

*The Myometrial
Smooth muscle
cells increase
in number and
size*

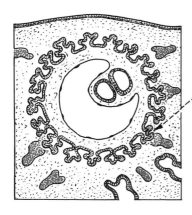

*CHORIONIC VILLI
– Finger-like
projections from
the embryo have
invaded
mother's
endometrial
blood vessels*

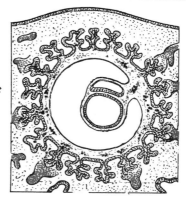

*Blood vessels
develop in the
CHORIONIC VILLI
which are now
interlocked with
mother's tissues
and surrounded
by mother's
blood.
STRUCTURE formed
in this way is
the PLACENTA.*

PLACENTA

The PLACENTA in WOMAN is HAEMOCHORIAL – *i.e. chorionic villi dip directly into maternal blood.*
It increases in weight throughout Pregnancy.

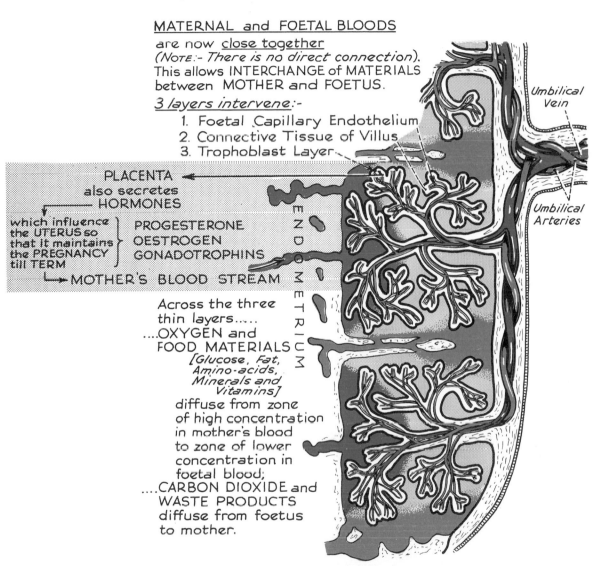

MATERNAL and FOETAL BLOODS
are now <u>close together</u>
(NOTE:- *There is no direct connection*).
This allows INTERCHANGE of MATERIALS
between MOTHER and FOETUS.

3 layers intervene:-
   1. Foetal Capillary Endothelium
   2. Connective Tissue of Villus
   3. Trophoblast Layer

PLACENTA
   also secretes
      HORMONES

which influence
the UTERUS so
that it maintains
the PREGNANCY
till TERM

PROGESTERONE
OESTROGEN
GONADOTROPHINS

→ MOTHER'S BLOOD STREAM

ENDOMETRIUM

Across the three
thin layers......
....OXYGEN and
FOOD MATERIALS
*[Glucose, Fat,
Amino-acids,
Minerals and
Vitamins]*
diffuse from zone
of high concentration
in mother's blood
to zone of lower
concentration in
foetal blood;
....CARBON DIOXIDE and
WASTE PRODUCTS
diffuse from foetus
to mother.

Umbilical
Vein

Umbilical
Arteries

*Substances of small molecular weight usually pass in either direction by diffusion.
Larger molecules are probably transported by special carrier systems involving enzymes.*

UTERUS

As <u>PREGNANCY ADVANCES</u>

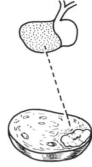

Foetus grows larger and comes to fill UTERINE CAVITY

Corpus Luteum ceases to function after 3RD or 4TH month.

Foetus is attached by UMBILICAL CORD to PLACENTA

Foetus is bathed by AMNIOTIC FLUID (within Amniotic and Chorionic membranes)

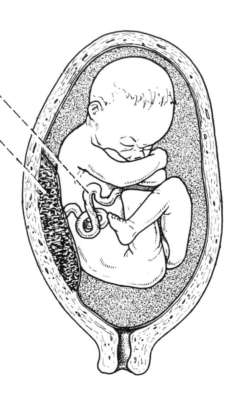

GROWTH of MYOMETRIUM —increase in number and size of smooth muscle cells and of the blood vessels

PLACENTAL OESTROGEN and PROGESTERONE GONADOTROPHINS in blood stream condition the maintenance of pregnancy

STRETCHING of MYOMETRIUM

192

FOETAL CIRCULATION

For the foetus the PLACENTA acts as the organ of transfer for Oxygen, Nutritives and Waste Products.   Only a small volume of blood passes through the foetal lungs.

BLOOD RETURNING TO HEART

...To *RIGHT ATRIUM*

Small amount from heart, head, neck and arms →S.V.C.

LARGE AMOUNT via UMBILICAL VEINS

through LIVER – short circuits to I.V.C. via **DUCTUS VENOSUS**.

Some of this passes to Right Atrium
Small amount from abdominal cavity and legs.

...To *LEFT ATRIUM*

Small amount from 2 lungs

LARGE AMOUNT from INFERIOR VENA CAVA through

FORAMEN OVALE.

(thus by-passing pulmonary circulation)

BLOOD LEAVING HEART

... From *RIGHT VENTRICLE*

Small amount to 2 lungs

LARGE AMOUNT to AORTA

through

DUCTUS ARTERIOSUS

(thus by-passing pulmonary circulation)
joins

OUTPUT from LEFT VENTRICLE

↓

Small amount to heart, head, neck and arms.
LARGE AMOUNT to PLACENTA
through
UMBILICAL ARTERIES
Small amount to abdominal cavity and legs.

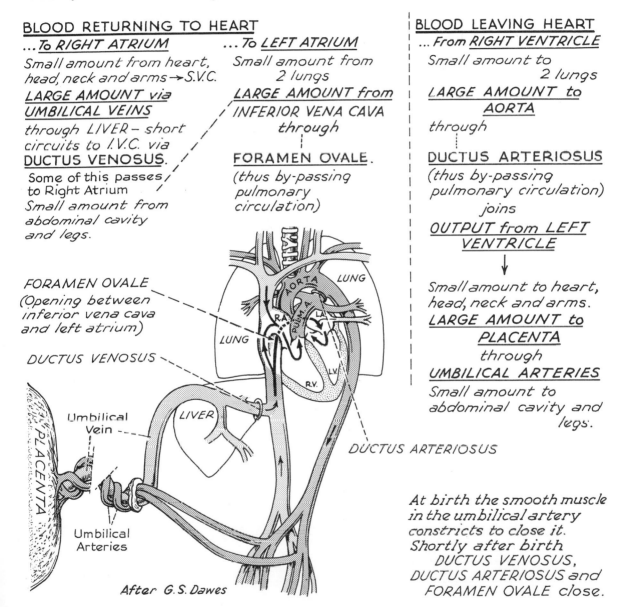

FORAMEN OVALE
(Opening between inferior vena cava and left atrium)

DUCTUS VENOSUS

Umbilical Vein

LIVER

Umbilical Arteries

DUCTUS ARTERIOSUS

After G. S. Dawes

At birth the smooth muscle in the umbilical artery constricts to close it. Shortly after birth DUCTUS VENOSUS, DUCTUS ARTERIOSUS and FORAMEN OVALE close.

UTERUS

PARTURITION

About 40 weeks after conception the process of <u>CHILDBIRTH</u> usually begins —— Factors initiating this are largely unknown.

After 32ND week of Pregnancy the uterus becomes sensitive to OXYTOCIN

1ST Stage usually lasts up to 14 hours with a first birth

2ND Stage LABOUR usually lasts up to 2 hours with a first birth

MYOMETRIUM

Uterine muscle is now very greatly stretched.

↓

Rhythmic contractions which begin to increase in strength and frequency

Press on Amniotic Fluid

Uterine contractions increase in strength and frequency (aided by voluntary contractions of Abdominal muscles)

↓

CHILD is slowly forced through PELVIS

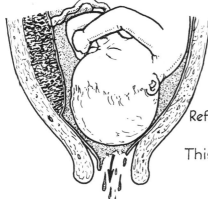

Neck of Womb (Cervix) and its opening (Os) dilate

Baby's head stretches CERVIX

Reflex OXYTOCIN secretion

This excites uterine contractions

These push down baby

Cervix further stretched

Stronger and stronger uterine contractions

Cycle repeats till baby is delivered

Membranes rupture and <u>AMNIOTIC FLUID</u> escapes

BIRTH of BABY

UTERUS

After PARTURITION

3RD Stage Labour
5-15 minutes
after birth of child

In PUERPERIUM

HYPOTHALAMUS

Afferents from Breasts in Suckling and from the uterus and birth canal during parturition

P.P.

Immediately
following childbirth.

FALL in
BLOOD LEVELS
of
OESTROGEN,
PROGESTERONE
and
GONADOTROPHINS
(after loss of Placenta)

OXYTOCIN
released to blood stream
stimulates
CONTRACTIONS of UTERINE
MUSCLE

↓

DETACH and DELIVER PLACENTA
and the membranes as the
AFTERBIRTH.

OXYTOCIN
released to
blood stream

MYOMETRIUM

Uterine muscle contracts down to
close off blood vessels torn and
bleeding after separation of
Placenta.

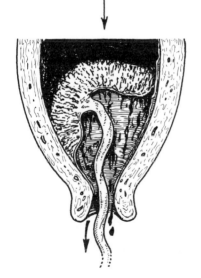

A large part of Endometrium —
Decidua — is shed with the
Placenta.
Only the limiting layer is left.

UTERUS

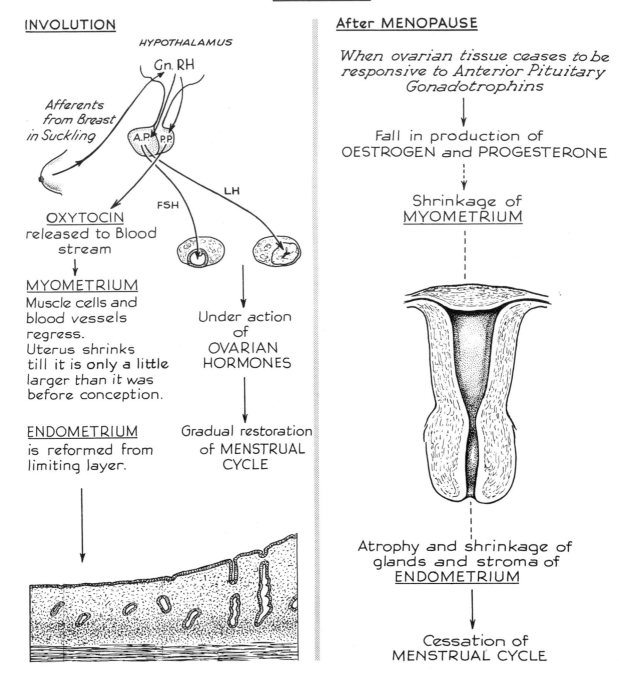

INVOLUTION

HYPOTHALAMUS

Gn. RH

Afferents from Breast in Suckling

A.P. P.P.

LH

FSH

OXYTOCIN
released to Blood stream

MYOMETRIUM
Muscle cells and blood vessels regress.
Uterus shrinks till it is only a little larger than it was before conception.

ENDOMETRIUM
is reformed from limiting layer.

Under action of OVARIAN HORMONES

Gradual restoration of MENSTRUAL CYCLE

After MENOPAUSE

When ovarian tissue ceases to be responsive to Anterior Pituitary Gonadotrophins

Fall in production of OESTROGEN and PROGESTERONE

Shrinkage of MYOMETRIUM

Atrophy and shrinkage of glands and stroma of ENDOMETRIUM

Cessation of MENSTRUAL CYCLE

MAMMARY GLANDS

There are 2 mammary glands.

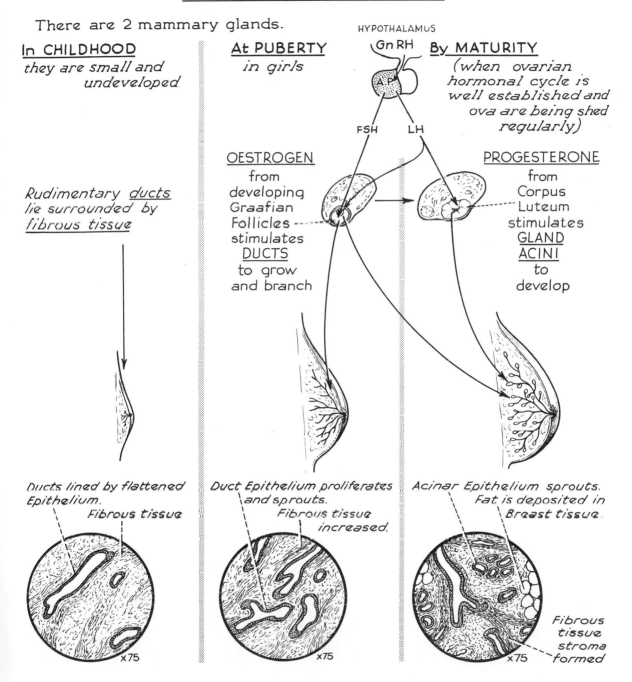

In CHILDHOOD
they are small and undeveloped

HYPOTHALAMUS
Gn RH

At PUBERTY
in girls

By MATURITY
(when ovarian hormonal cycle is well established and ova are being shed regularly)

FSH   LH

Rudimentary ducts lie surrounded by fibrous tissue

OESTROGEN
from developing Graafian Follicles stimulates DUCTS to grow and branch

PROGESTERONE
from Corpus Luteum stimulates GLAND ACINI to develop

Ducts lined by flattened Epithelium.
Fibrous tissue

Duct Epithelium proliferates and sprouts.
Fibrous tissue increased.

Acinar Epithelium sprouts.
Fat is deposited in Breast tissue.

Fibrous tissue stroma formed

×75   ×75   ×75

MAMMARY GLANDS

In PREGNANCY

PROGESTERONE
from
Corpus Luteum
and
OESTROGEN and
PROGESTERONE
from Placenta
stimulate
further GROWTH
of DUCTS and
ACINI.

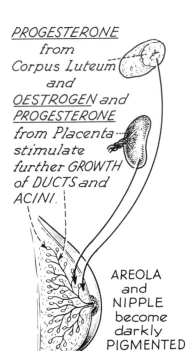

AREOLA
and
NIPPLE
become
darkly
PIGMENTED

_Acinar Epithelium becomes
cuboidal_

×75

After CHILDBIRTH

⟐ _HYPOTHALAMUS_

Fall in OESTROGEN
and PROGESTERONE
(after loss of Placenta)
inhibits secretion of
HYPOTHALAMIC-
PROLACTIN-INHIBITING-
FACTOR.

PROLACTIN is
released by
ANTERIOR
PITUITARY

stimulates
prepared
GLAND ACINI
to secrete
MILK

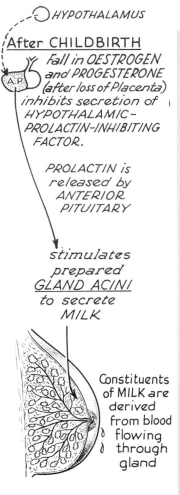

Constituents
of MILK are
derived
from blood
flowing
through
gland

Apocrine Acini – luminal
margin
breaks
down
to
release
secretion

×75

Emotional factors influence LACTATION

Milk production
starts 3-4 days
after childbirth
and is maintained
(often for several
months) by
PROLACTIN.

AFFERENT nerve
impulses, set up
by Child _SUCKLING_
give rise to reflex
EFFERENT nerve
impulses to Post.
Pituitary for the
release of _OXYTOCIN_
-carried by blood
stream to stimulate
the 'let-down' of
milk which is
then more readily
available
to the
suckling
child.

(see p.169)

HYPO-
THALAMUS

PROLACTIN

OXYTOCIN

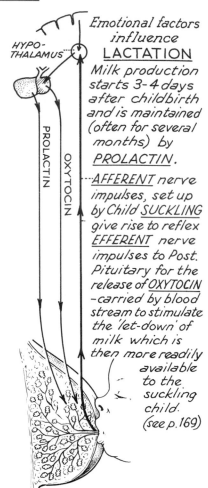

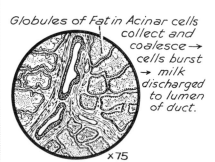

_Globules of Fat in Acinar cells
collect and
coalesce →
cells burst
→ milk
discharged
to lumen
of duct._

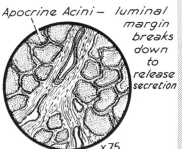

×75

MAMMARY GLANDS

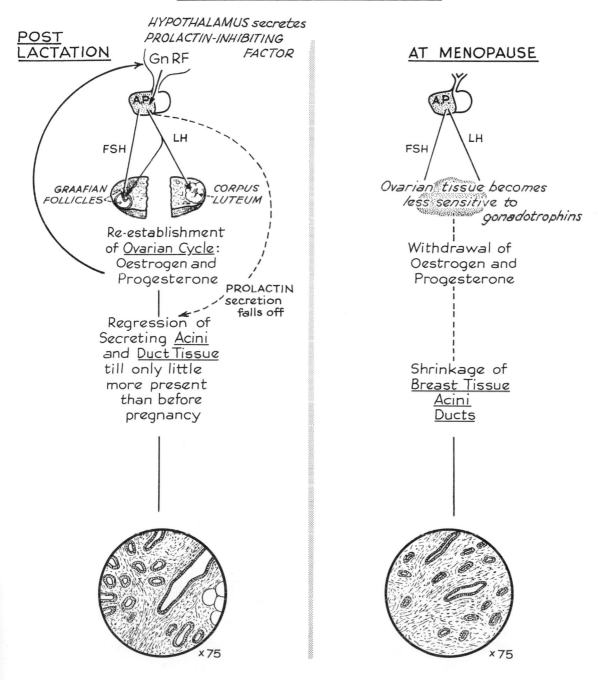

POST LACTATION

HYPOTHALAMUS secretes PROLACTIN-INHIBITING FACTOR

Gn RF

A.P.

FSH   LH

GRAAFIAN FOLLICLES   CORPUS LUTEUM

Re-establishment of Ovarian Cycle: Oestrogen and Progesterone

PROLACTIN secretion falls off

Regression of Secreting Acini and Duct Tissue till only little more present than before pregnancy

×75

AT MENOPAUSE

A.P.

FSH   LH

Ovarian tissue becomes less sensitive to gonadotrophins

Withdrawal of Oestrogen and Progesterone

Shrinkage of Breast Tissue Acini Ducts

×75

MENOPAUSE

Between the ages of 42 and 50 years OVARIAN tissue gradually ceases to respond to stimulation by <u>ANTERIOR PITUITARY GONADO-TROPHIC HORMONES.</u>

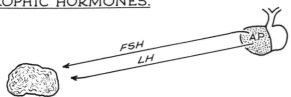

OVARIAN CYCLE becomes irregular and finally ceases ⟶ Ovary becomes small and fibrosed and no longer produces ripe Ova.

OESTROGEN and PROGESTERONE levels in Blood stream fall.

TISSUES of the body ⟶ begin to show changes which mark the end of REPRODUCTIVE LIFE.

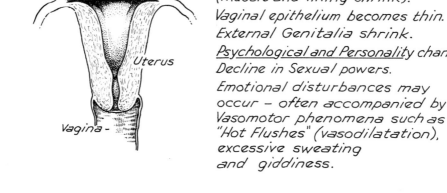

Sometimes final redistribution of fat → less typically feminine distribution.

Regression of <u>Secondary Sex Characteristics</u>.

Breasts shrink.

Hair becomes sparse in axillae and on pubis.

<u>*Secondary Sex Organs*</u> *atrophy.*

Fallopian tubes shrink.

Uterine Cycle and Menstruation cease.

(Muscle and lining shrink).

Vaginal epithelium becomes thin.

External Genitalia shrink.

<u>*Psychological and Personality*</u> *changes.*

Decline in Sexual powers.

Emotional disturbances may occur – often accompanied by Vasomotor phenomena such as "Hot Flushes" (vasodilatation), excessive sweating and giddiness.

After the MENOPAUSE a woman is usually unable to bear children.

PITUITARY, OVARIAN and ENDOMETRIAL CYCLES

HYPOTHALAMUS secretes Gonadotrophin Releasing Factor into hypothalamic-hypophyseal portal circulation (page 184).

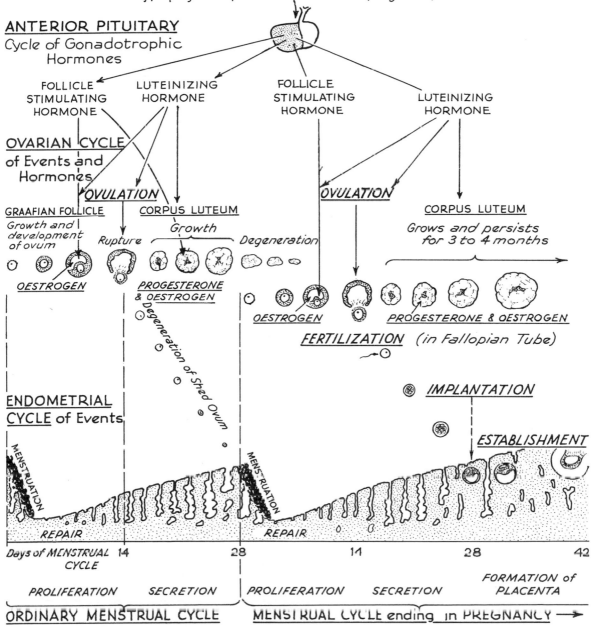

Days of MENSTRUAL CYCLE 14   28   14   28   42

ORDINARY MENSTRUAL CYCLE   MENSTRUAL CYCLE ending in PREGNANCY →

CHAPTER 9

"MASTER" TISSUES:

CENTRAL NERVOUS SYSTEM
LOCOMOTOR SYSTEM

NERVOUS SYSTEM

The Nervous System is concerned with the INTEGRATION and CONTROL of all bodily functions.

It has specialized in IRRITABILITY — *the ability to receive and respond to messages from the external and internal environments*

and also in CONDUCTION – *the ability to transmit messages to and from CO-ORDINATING CENTRES.*

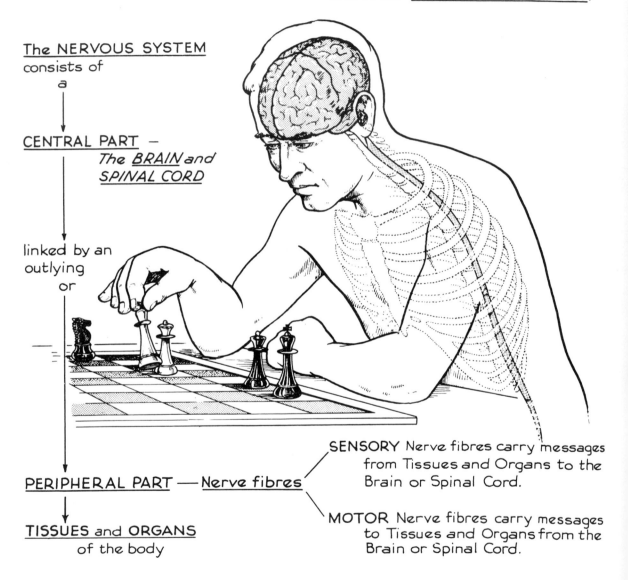

The NERVOUS SYSTEM consists of
a
↓

CENTRAL PART –
The BRAIN and SPINAL CORD

↓

linked by an outlying or

↓

PERIPHERAL PART —— Nerve fibres

↓

TISSUES and ORGANS
of the body

SENSORY Nerve fibres carry messages from Tissues and Organs to the Brain or Spinal Cord.

MOTOR Nerve fibres carry messages to Tissues and Organs from the Brain or Spinal Cord.

DEVELOPMENT of the NERVOUS SYSTEM

The Nervous System develops in the embryo from a simple tube of ECTODERM:- The PRIMITIVE NEURAL TUBE.

The CELLS lining it become the NERVOUS tissue of the BRAIN and SPINAL CORD. The CANAL becomes distended to form the VENTRICLES of the BRAIN and CENTRAL CANAL of the SPINAL CORD:-

FORE MID HIND

Prosencephalon   Mesencephalon   Rhombencephalon

Each of these swellings with the cells lining them become more complicated:-

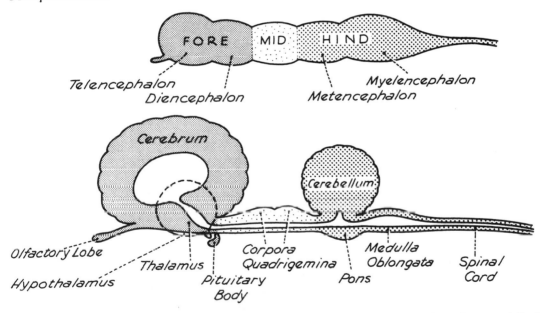

Telencephalon
Diencephalon

Myelencephalon
Metencephalon

Cerebrum

Cerebellum

Olfactory Lobe
Thalamus
Hypothalamus
Pituitary Body
Corpora Quadrigemina
Pons
Medulla Oblongata
Spinal Cord

There are now many layers of cells forming the Brain and Spinal Cord. The Ventricles of the Brain and the Central Canal of the Spinal Cord are filled with CEREBRO-SPINAL FLUID.

CEREBRUM

The largest part of the human brain is the CEREBRUM — made up of 2 CEREBRAL HEMISPHERES. Each of these is divided into LOBES.

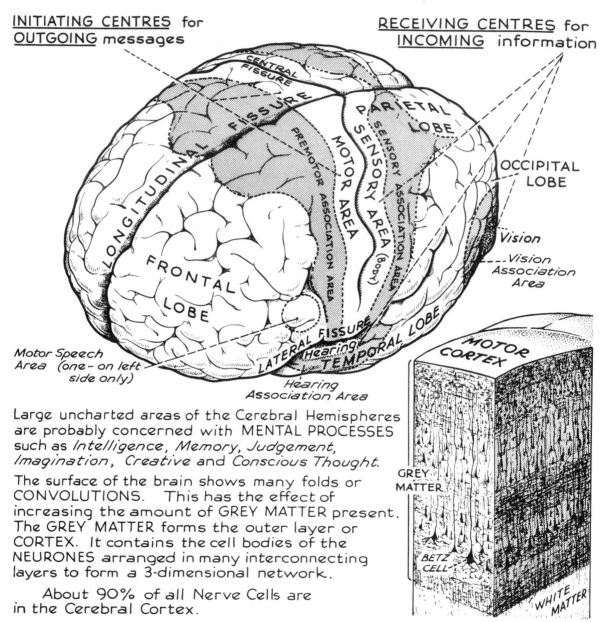

INITIATING CENTRES for OUTGOING messages

RECEIVING CENTRES for INCOMING information

CENTRAL FISSURE

PARIETAL LOBE

LONGITUDINAL FISSURE

PREMOTOR ASSOCIATION AREA

MOTOR AREA

SENSORY AREA (Body)

SENSORY ASSOCIATION AREA

OCCIPITAL LOBE

Vision

Vision Association Area

FRONTAL LOBE

LATERAL FISSURE

Hearing

TEMPORAL LOBE

Motor Speech Area (one - on left side only)

Hearing Association Area

MOTOR CORTEX

GREY MATTER

BETZ CELL

WHITE MATTER

Large uncharted areas of the Cerebral Hemispheres are probably concerned with MENTAL PROCESSES such as *Intelligence, Memory, Judgement, Imagination, Creative* and *Conscious Thought.*

The surface of the brain shows many folds or CONVOLUTIONS. This has the effect of increasing the amount of GREY MATTER present. The GREY MATTER forms the outer layer or CORTEX. It contains the cell bodies of the NEURONES arranged in many interconnecting layers to form a 3-dimensional network.

About 90% of all Nerve Cells are in the Cerebral Cortex.

HORIZONTAL SECTION through BRAIN

This view shows surface GREY MATTER containing Nerve Cells and inner WHITE MATTER made up of Nerve Fibres.

Deep in the substance of the Cerebral Hemispheres there are additional masses of GREY MATTER :-

The BASAL GANGLIA   *and* THALAMUS

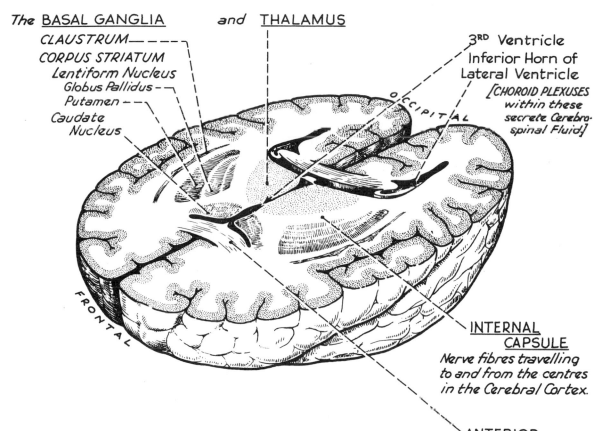

CLAUSTRUM _ _ _ _
CORPUS STRIATUM
Lentiform Nucleus
Globus Pallidus _ _
Putamen _ _
Caudate
Nucleus

OCCIPITAL

FRONTAL

3RD Ventricle
Inferior Horn of
Lateral Ventricle
[CHOROID PLEXUSES
within these
secrete Cerebro-
spinal Fluid]

INTERNAL
CAPSULE
Nerve fibres travelling
to and from the centres
in the Cerebral Cortex.

ANTERIOR
COMMISSURE
Nerve fibres linking
the two hemispheres.

The Basal Ganglia
are concerned with
modifying and
co-ordinating
VOLUNTARY MUSCLE
MOVEMENT.

The Thalamus
is an important
relay centre for
SENSORY fibres
on their way to
CEREBRAL CORTEX.
'Crude' Sensation
and PAIN may be
appreciated here.

207

VERTICAL SECTION through BRAIN

This is a Vertical Section through the LONGITUDINAL FISSURE — a deep cleft which separates the two Cerebral Hemispheres. At the bottom of this cleft are tracts of nerve fibres which link up the different LOBES of each hemisphere and also link the two hemispheres with each other — the CORPUS CALLOSUM.

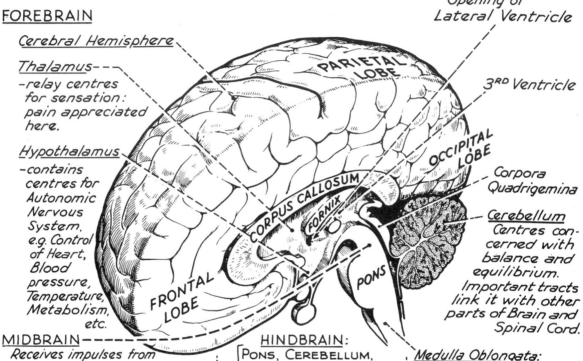

FOREBRAIN

Cerebral Hemisphere

Thalamus—
-relay centres for sensation: pain appreciated here.

Hypothalamus
-contains centres for Autonomic Nervous System. e.g. Control of Heart, Blood pressure, Temperature, Metabolism, etc.

Opening of Lateral Ventricle

3RD Ventricle

Corpora Quadrigemina

Cerebellum
Centres concerned with balance and equilibrium. Important tracts link it with other parts of Brain and Spinal Cord.

Labels on figure: PARIETAL LOBE, OCCIPITAL LOBE, CORPUS CALLOSUM, FORNIX, FRONTAL LOBE, PONS

MIDBRAIN
Receives impulses from Retina and Ear. Serves as a centre for Visual and Auditory Reflexes. In the <u>Grey Matter</u> are nerve cell bodies of III, IV Cranial nerves and the Red Nucleus which helps to control skilled muscular movements. The <u>White Matter</u> carries nerve fibres linking Red Nucleus with Cerebral Cortex, Thalamus, Cerebellum, Corpus Striatum and Spinal Cord. It also carries Ascending Sensory fibres in Lateral and Medial Lemnisci, and Descending Motor fibres on their way to Pons and Spinal Cord.

HINDBRAIN:
[PONS, CEREBELLUM, MEDULLA OBLONGATA]

<u>Pons</u>: Groups of Neurones form sensory nucleus of V and also nuclei of VI and VII Cranial nerves. Other nerve cells here relay impulses along their axons to Cerebellum and Cerebrum. Rubrospinal tract, Lateral and Medial Lemnisci pass through Pons and nerve fibres linking Cerebral Cortex with Medulla Oblongata and Spinal Cord.

<u>Medulla Oblongata</u>:
Groups of Neurones form Nuclei of VIII, IX, X, XI, XII Cranial nerves. Gracile and Cuneate Nuclei -second sensory neurones in cutaneous pathways. Tracts of Sensory fibres decussate and ascend to other side of Cerebral Cortex. Some fibres remain uncrossed. The larger part of each Motor pyramidal tract crosses and descends in other side of Spinal Cord.

CORONAL SECTION through BRAIN

This is a section through the TRANSVERSE (CENTRAL) FISSURE.
It shows each of the major developments of the BRAIN –

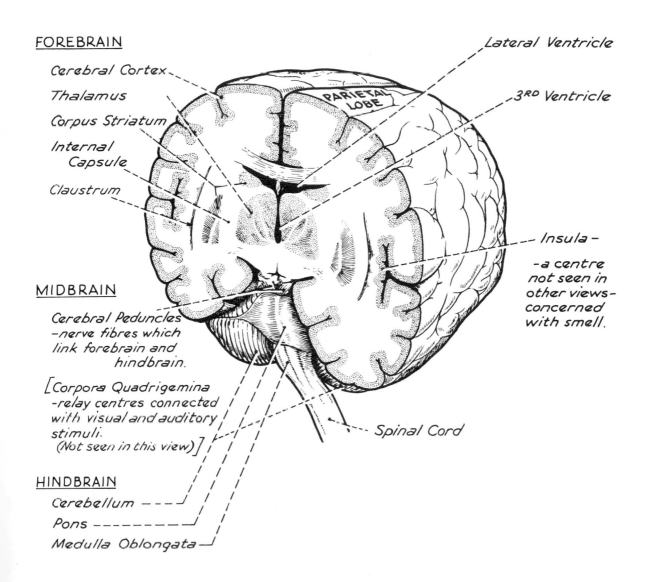

FOREBRAIN

Cerebral Cortex

Thalamus

Corpus Striatum

Internal
Capsule

Claustrum

PARIETAL
LOBE

Lateral Ventricle

3ᴿᴰ Ventricle

Insula –

– a centre
not seen in
other views –
concerned
with smell.

MIDBRAIN

Cerebral Peduncles
– nerve fibres which
link forebrain and
hindbrain.

[Corpora Quadrigemina
– relay centres connected
with visual and auditory
stimuli.
(Not seen in this view)]

Spinal Cord

HINDBRAIN

Cerebellum – – – –

Pons – – – – – – – –

Medulla Oblongata

CRANIAL NERVES

Twelve pairs of nerves arise directly from the undersurface of the Brain to supply Head and Neck and most of the viscera.

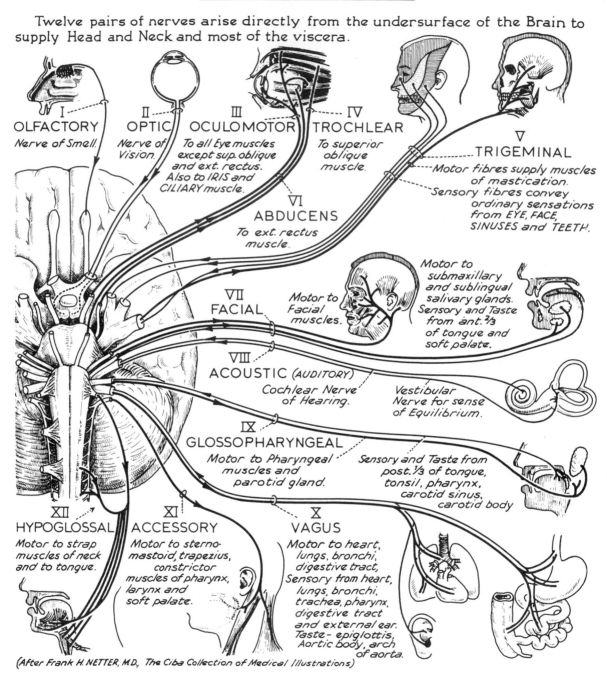

I OLFACTORY
Nerve of Smell.

II OPTIC
Nerve of Vision.

III OCULOMOTOR
To all Eye muscles except sup. oblique and ext. rectus. Also to IRIS and CILIARY muscle.

IV TROCHLEAR
To superior oblique muscle.

V TRIGEMINAL
Motor fibres supply muscles of mastication.
Sensory fibres convey ordinary sensations from EYE, FACE, SINUSES and TEETH.

VI ABDUCENS
To ext. rectus muscle.

VII FACIAL
Motor to Facial muscles.
Motor to submaxillary and sublingual salivary glands. Sensory and Taste from ant. 2/3 of tongue and soft palate.

VIII ACOUSTIC (AUDITORY)
Cochlear Nerve of Hearing.
Vestibular Nerve for sense of Equilibrium.

IX GLOSSOPHARYNGEAL
Motor to Pharyngeal muscles and parotid gland.
Sensory and Taste from post. 1/3 of tongue, tonsil, pharynx, carotid sinus, carotid body

XII HYPOGLOSSAL
Motor to strap muscles of neck and to tongue.

XI ACCESSORY
Motor to sterno-mastoid, trapezius, constrictor muscles of pharynx, larynx and soft palate.

X VAGUS
Motor to heart, lungs, bronchi, digestive tract, Sensory from heart, lungs, bronchi, trachea, pharynx, digestive tract and external ear. Taste - epiglottis, Aortic body, arch of aorta.

(After Frank H. NETTER, M.D., The Ciba Collection of Medical Illustrations)

210

SPINAL CORD

The SPINAL CORD lies within the Vertebral Canal. It is continuous above with the Medulla Oblongata.

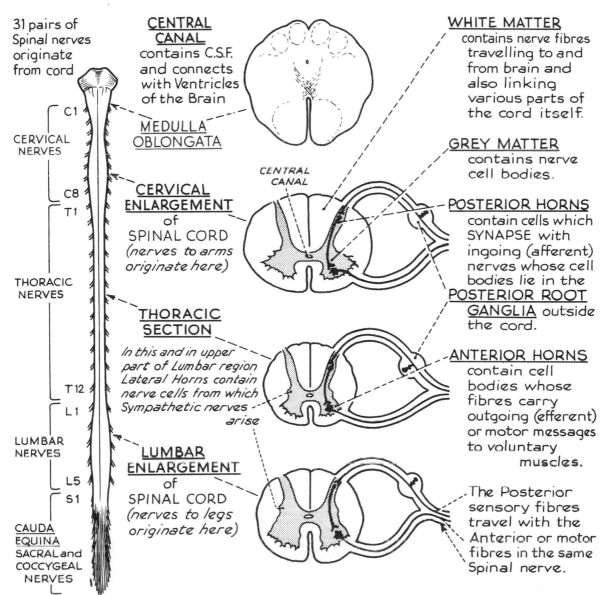

31 pairs of Spinal nerves originate from cord

CERVICAL NERVES

THORACIC NERVES

LUMBAR NERVES

CAUDA EQUINA SACRAL and COCCYGEAL NERVES

C1

C8
T1

T 12
L1

L5
S1

CENTRAL CANAL contains C.S.F. and connects with Ventricles of the Brain

MEDULLA OBLONGATA

CENTRAL CANAL

CERVICAL ENLARGEMENT of SPINAL CORD *(nerves to arms originate here)*

THORACIC SECTION

In this and in upper part of Lumbar region Lateral Horns contain nerve cells from which Sympathetic nerves arise

LUMBAR ENLARGEMENT of SPINAL CORD *(nerves to legs originate here)*

WHITE MATTER contains nerve fibres travelling to and from brain and also linking various parts of the cord itself.

GREY MATTER contains nerve cell bodies.

POSTERIOR HORNS contain cells which SYNAPSE with ingoing (afferent) nerves whose cell bodies lie in the **POSTERIOR ROOT GANGLIA** outside the cord.

ANTERIOR HORNS contain cell bodies whose fibres carry outgoing (efferent) or motor messages to voluntary muscles.

The Posterior sensory fibres travel with the Anterior or motor fibres in the same Spinal nerve.

The spinal nerves travel to all parts of the trunk and limbs.

211

SYNAPSE

The structural unit of the Nervous System is the NEURONE.
Neurones are linked together in the Nervous System.....

..... Nerve Process to Nerve Cell body at a SYNAPSE

or

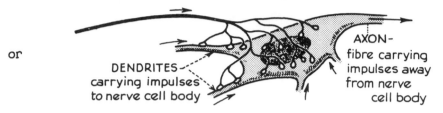

DENDRITES -
carrying impulses
to nerve cell body

AXON -
fibre carrying
impulses away
from nerve
cell body

*AXON ends in small
swellings – END FEET -
which merely touch
the DENDRITES or
BODY of another
NERVE CELL.
i.e. There is no direct
protoplasmic union
between neurones
at the SYNAPSE.*

..... Nerve Process to Nerve Process at a SYNAPSE

*One neurone usually connects with a great many others often widely
scattered in different parts of the Brain and Spinal Cord. In this way intricate
chains of nerve cells forming complex pathways for incoming and outgoing
information can be built up within the Central Nervous System.*

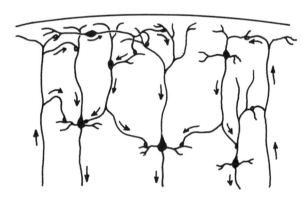

*When the NERVE IMPULSE – like a very small electrical current — reaches a
SYNAPSE it causes the release at the NERVE ENDINGS of a CHEMICAL SUBSTANCE
which bridges the gap and forms the stimulus to conduct the Impulse to
the next Neurone.*

A Synapse permits transmission of the impulse in one direction only.

NERVE IMPULSE

Neurones specialize in IRRITABILITY and in CONDUCTION of IMPULSES.

In *RESTING (INACTIVE) NERVE FIBRE*:

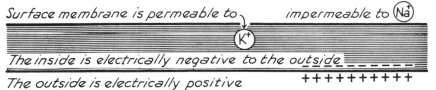

Surface membrane is permeable to   *impermeable to* Na^+

The inside is electrically negative to the outside

The outside is electrically positive $+ + + + + + + + +$

This is the so-called 'POLARIZED' or resting STATE. It is the result of chemical processes in the cells and is a potential source of energy.

When *NERVE FIBRE is STIMULATED* (e.g. electrically or chemically) ELECTRO-CHEMICAL CHANGES occur at point of stimulation. The resting potential is abolished.

The membrane is said to be 'DEPOLARIZED'.

The membrane permeability is altered

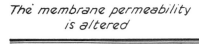

The electrical state is reversed. The inside becomes electrically positive and the outside becomes electrically negative.

Current flows ahead from the stimulated or active part to the inactive part – carried largely by the Potassium and Sodium ions.

Current passing through the membrane of the inactive part in turn depolarizes it. This then becomes the new active area to excite the next segment.

Direction of Impulse ⟶

In *MYELINATED NERVE FIBRE*:

The active region still 'triggers' the resting part ahead of it by causing an outward flow of current but this occurs only at the Nodes of Ranvier. i.e. The Nerve Impulse is propagated from Node to Node.

This increases speed of CONDUCTION along these large fibres.

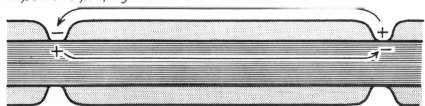

After passage of nerve impulse 'REPOLARIZATION' and IONIC BALANCE are quickly restored.

A nerve impulse is the same whether it is in a sensory nerve taking messages to the brain or in a motor nerve taking impulses from the brain to the muscles, etc.

REFLEX ACTION

The NEURONE is the ANATOMICAL or STRUCTURAL UNIT of the Nervous System: the Nervous REFLEX is the PHYSIOLOGICAL or FUNCTIONAL UNIT.

A *Nervous Reflex* is an INVOLUNTARY ACTION caused by the STIMULATION of an AFFERENT (*sensory*) nerve ending or RECEPTOR.

The structural basis of Reflex Action is the REFLEX ARC. In its simplest form this consists of :-

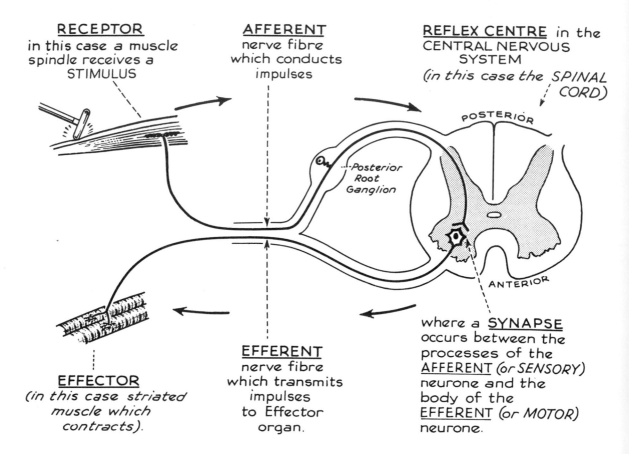

RECEPTOR
in this case a muscle spindle receives a STIMULUS

AFFERENT
nerve fibre which conducts impulses

REFLEX CENTRE in the CENTRAL NERVOUS SYSTEM
(*in this case the SPINAL CORD*)

POSTERIOR

Posterior Root Ganglion

ANTERIOR

EFFECTOR
(*in this case striated muscle which contracts*).

EFFERENT
nerve fibre which transmits impulses to Effector organ.

where a **SYNAPSE** occurs between the processes of the **AFFERENT** (*or SENSORY*) neurone and the body of the **EFFERENT** (*or MOTOR*) neurone.

Reflexes form the basis of all Central Nervous System (C.N.S.) activity. They occur at all levels of the Brain and Spinal Cord. Important bodily functions such as movements of Respiration, Digestion, etc, are all controlled through Reflexes. We are made aware of some reflex acts: others occur without our knowledge.

214

STRETCH REFLEXES

In man a very few REFLEX ARCS involve 2 NEURONES only.
Two examples elicited by doctors when testing the Nervous System
are:—

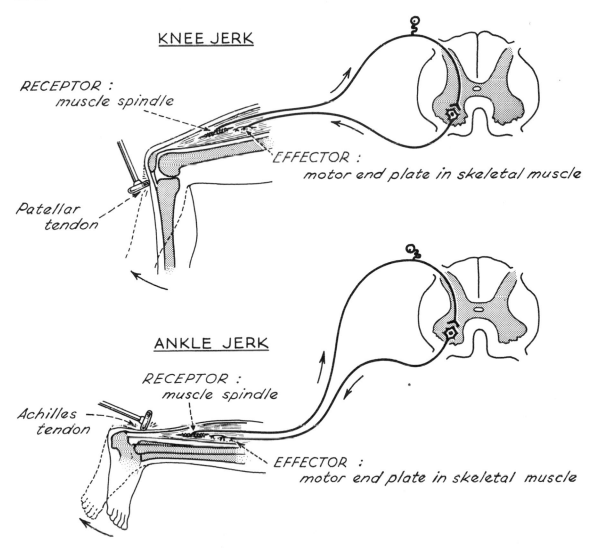

KNEE JERK

RECEPTOR :
muscle spindle

EFFECTOR :
motor end plate in skeletal muscle

Patellar
tendon

ANKLE JERK

RECEPTOR :
muscle spindle

Achilles
tendon

EFFECTOR :
motor end plate in skeletal muscle

When the tendon is sharply tapped the muscle is stretched. (N.B. The
stimulus is by *stretch* of the *muscle spindle*.) Nerve messages pass
into the spinal cord — and out to the muscle which then contracts.

SPINAL REFLEXES

In most REFLEX ARCS in man AFFERENT and EFFERENT NEURONES are linked by at least 1 CONNECTOR NEURONE.

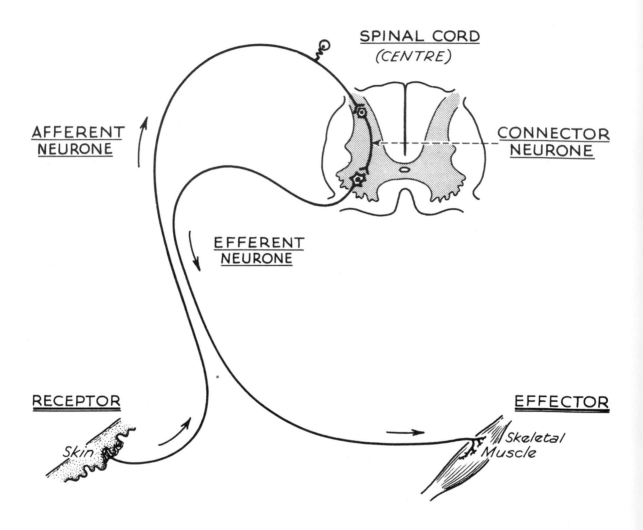

SPINAL CORD
(CENTRE)

AFFERENT NEURONE

CONNECTOR NEURONE

EFFERENT NEURONE

RECEPTOR

Skin

EFFECTOR

Skeletal Muscle

A chain of many connector neurones is frequently found.

"EDIFICE" of the C.N.S. (After R.C. GARRY)

In the majority of Reflex arcs in man a chain of many connector neurones is found. There may be link-ups with various levels of the Brain and Spinal Cord.

This diagram gives a highly simplified concept of the type of LINK-UP which can occur between different levels of the Central Nervous System.

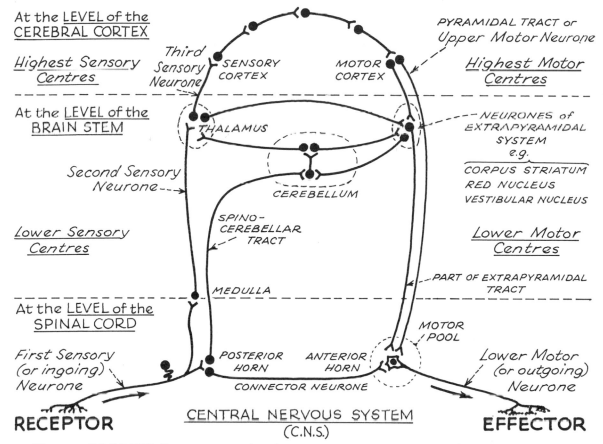

At the LEVEL of the CEREBRAL CORTEX

Highest Sensory Centres

Third Sensory Neurone

SENSORY CORTEX

MOTOR CORTEX

PYRAMIDAL TRACT or Upper Motor Neurone

Highest Motor Centres

At the LEVEL of the BRAIN STEM

THALAMUS

NEURONES of EXTRAPYRAMIDAL SYSTEM e.g.

CORPUS STRIATUM
RED NUCLEUS
VESTIBULAR NUCLEUS

Second Sensory Neurone

CEREBELLUM

Lower Sensory Centres

SPINO-CEREBELLAR TRACT

Lower Motor Centres

MEDULLA

PART OF EXTRAPYRAMIDAL TRACT

At the LEVEL of the SPINAL CORD

MOTOR POOL

First Sensory (or ingoing) Neurone

POSTERIOR HORN

ANTERIOR HORN

Lower Motor (or outgoing) Neurone

CONNECTOR NEURONE

RECEPTOR

CENTRAL NERVOUS SYSTEM (C.N.S.)

EFFECTOR

Every RECEPTOR neurone is thus potentially linked in the C.N.S. with a large number of EFFECTOR organs all over the body and every EFFECTOR neurone is similarly in communication with RECEPTORS all over the body.

Centres in the BRAIN and BRAIN STEM can thus modify REFLEX ACTS which occur through the SPINAL CORD. These centres can send "suppressing" or "facilitating" impulses along their pathways to the cells in the SPINAL CORD.

REFLEX ACTION

Most REFLEX ACTIONS in man involve a great many REFLEX ARCS.

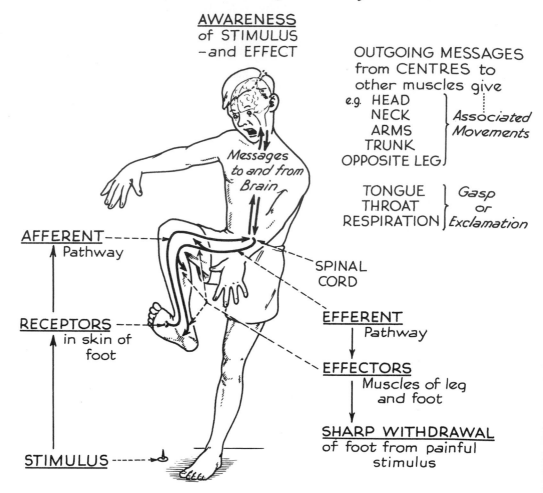

AWARENESS
of STIMULUS
–and EFFECT

OUTGOING MESSAGES
from CENTRES to
other muscles give
e.g. HEAD
NECK
ARMS   } *Associated*
TRUNK   *Movements*
OPPOSITE LEG }

TONGUE   } *Gasp*
THROAT   *or*
RESPIRATION } *Exclamation*

*Messages
to and from
Brain*

AFFERENT
Pathway

SPINAL
CORD

EFFERENT
Pathway

RECEPTORS
in skin of
foot

EFFECTORS
Muscles of leg
and foot

STIMULUS

SHARP WITHDRAWAL
of foot from painful
stimulus

The localized stimulation of a very few RECEPTORS – sending messages along their AFFERENT NEURONES to SPINAL CORD and to BRAIN – has led to a very large number of outgoing impulses in many EFFECTOR NEURONES to a large number of MUSCLES to give a very widespread and generalized REFLEX RESPONSE.

This is possible because each receptor neurone is potentially connected within the Central Nervous System with all effector neurones.

ARRANGEMENT of NEURONES

Some of the ways in which neurones can be linked are indicated here:-

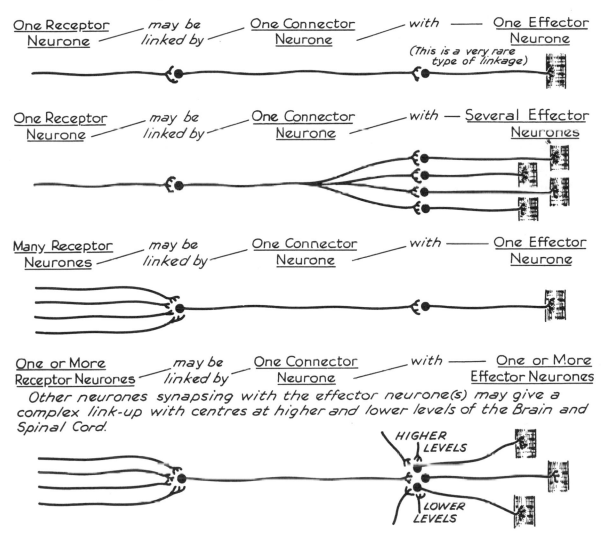

One Receptor Neurone — may be linked by — One Connector Neurone — with — One Effector Neurone (This is a very rare type of linkage)

One Receptor Neurone — may be linked by — One Connector Neurone — with — Several Effector Neurones

Many Receptor Neurones — may be linked by — One Connector Neurone — with — One Effector Neurone

One or More Receptor Neurones — may be linked by — One Connector Neurone — with — One or More Effector Neurones

Other neurones synapsing with the effector neurone(s) may give a complex link-up with centres at higher and lower levels of the Brain and Spinal Cord.

HIGHER LEVELS

LOWER LEVELS

Through such 'functional' link-ups, neurones in different parts of the Central Nervous System, when active, can influence each other. This makes it possible for 'CONDITIONED' REFLEXES to become established (for simple example see page 55).
Such reflexes probably form the basis of all training so that it becomes difficult to say where REFLEX (or INVOLUNTARY) behaviour ends and purely VOLUNTARY behaviour begins.

SENSE ORGANS

Man's awareness of the world is limited to those forms of energy, physical or chemical, to which he has Receptors designed to respond. (Many "events" in the universe go unnoticed by man because he has no Sense Organ which can respond to them.)

Each Sense Organ is designed to respond to one type of stimulation. EXTEROCEPTORS are stimulated by events in the EXTERNAL ENVIRONMENT.

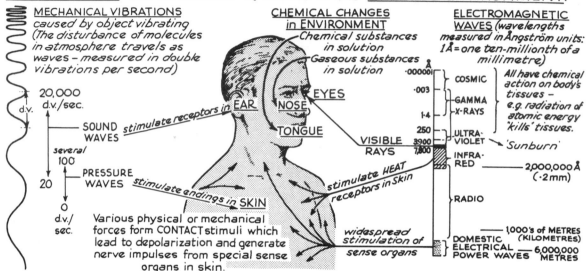

MECHANICAL VIBRATIONS caused by object vibrating (The disturbance of molecules in atmosphere travels as waves - measured in double vibrations per second)

20,000 d.v./sec.
d.v.
SOUND WAVES *stimulate receptors in* EAR
several 100
20
PRESSURE WAVES *stimulate endings in* SKIN
0 d.v./sec.

Various physical or mechanical forces form CONTACT stimuli which lead to depolarization and generate nerve impulses from special sense organs in skin.

CHEMICAL CHANGES in ENVIRONMENT
Chemical substances in solution
Gaseous substances in solution
EYES NOSE TONGUE
stimulate HEAT receptors in Skin
widespread stimulation of sense organs

ELECTROMAGNETIC WAVES (wavelengths measured in Ångström units: 1Å = one ten-millionth of a millimetre)

Å		
·00001	COSMIC	All have chemical action on body's tissues – e.g. radiation of atomic energy 'kills' tissues.
·003	GAMMA X-RAYS	
1·4		
250	ULTRA-VIOLET	'Sunburn'
3900 VISIBLE		
7800 RAYS	INFRA-RED	2,000,000 Å (·2mm)
	RADIO	
		1,000's of METRES (KILOMETRES)
	DOMESTIC ELECTRICAL POWER WAVES	6,000,000 METRES

Exteroceptors may convey information to CONSCIOUSNESS with AWARENESS or SENSATION and lead to suitable RESPONSES planned in CEREBRAL CORTEX or they may serve as AFFERENT pathways for REFLEX (or INVOLUNTARY) ACTION with or without rising to consciousness.

PROPRIOCEPTORS are stimulated by changes in LOCOMOTOR SYSTEM of body

LABYRINTH ----- *movements and position of head*
MUSCLES -------- *stretch*
TENDONS ------- *tension and stretch*
JOINTS ---------- *stretch and pressure*

} *SENSE of EQUILIBRIUM or BALANCE and AWARENESS of POSITION and MOVEMENT of BODY in space.*

INTEROCEPTORS in VISCERA are stimulated by changes in INTERNAL ENVIRONMENT (*e.g. by distension in hollow organs*).

Much of the PROPRIO- and INTEROCEPTOR information never rises to consciousness.

Overstimulation of any receptor can give rise to sensation of PAIN.

Most receptors show ADAPTATION — if continuously stimulated they send reduced numbers of impulses to the brain.

PERCEPTION is awareness of the source of the stimulus.
Information from one Sense Organ is correlated with other information (*past or present*).
APPERCEPTION is recognition or identification of the source of the stimulus and depends on association with past experience.

SMELL

Smell is a CHEMICAL SENSE, i.e. the receptors respond to CHEMICAL STIMULI. To arouse the sensation a substance must first be in a GASEOUS STATE then go into SOLUTION.

<u>The ORGAN of SMELL is the NOSE</u> — *Also serves as the main air passage to Respiratory System.*

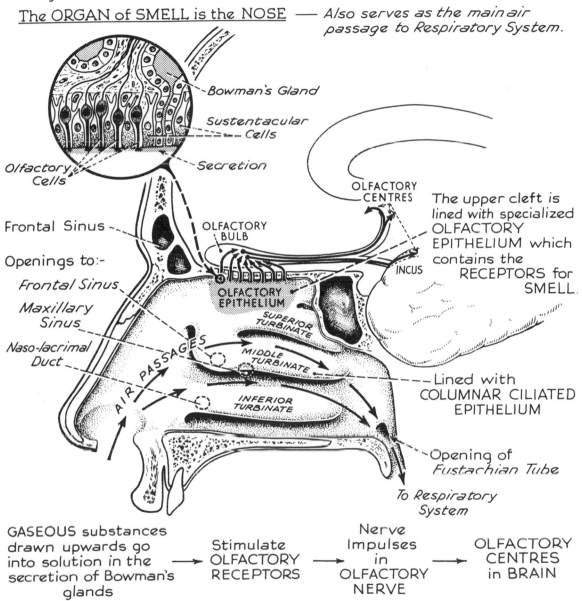

Bowman's Gland

Sustentacular Cells

Olfactory Cells

Secretion

Frontal Sinus

Openings to:-

Frontal Sinus

Maxillary Sinus

Naso-lacrimal Duct

OLFACTORY BULB

AIR PASSAGES

OLFACTORY EPITHELIUM

SUPERIOR TURBINATE

MIDDLE TURBINATE

INFERIOR TURBINATE

OLFACTORY CENTRES

INCUS

The upper cleft is lined with specialized OLFACTORY EPITHELIUM which contains the RECEPTORS for SMELL.

Lined with COLUMNAR CILIATED EPITHELIUM

Opening of *Eustachian Tube*

To Respiratory System

GASEOUS substances drawn upwards go into solution in the secretion of Bowman's glands ⟶ Stimulate OLFACTORY RECEPTORS ⟶ Nerve Impulses in OLFACTORY NERVE ⟶ OLFACTORY CENTRES in BRAIN

TASTE

Taste is a CHEMICAL SENSE, i.e. receptors respond to CHEMICAL STIMULI. To arouse the sensation a substance must be in SOLUTION.

The essential <u>ORGAN of TASTE</u> is the TONGUE —— The VOLUNTARY muscular organ concerned also in MASTICATION, SWALLOWING and SPEECH.

Covered with STRATIFIED SQUAMOUS EPITHELIUM.

Projections on its upper surface are called

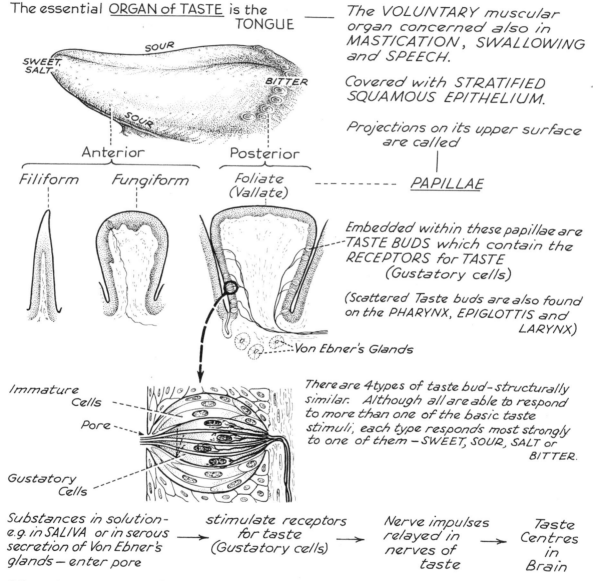

PAPILLAE

SOUR
SWEET, SALT
BITTER
SOUR

Anterior   Posterior

Filiform   Fungiform   Foliate (Vallate)

Embedded within these papillae are TASTE BUDS which contain the RECEPTORS for TASTE (Gustatory cells)

(Scattered Taste buds are also found on the PHARYNX, EPIGLOTTIS and LARYNX)

Von Ebner's Glands

Immature Cells
Pore
Gustatory Cells

There are 4 types of taste bud - structurally similar. Although all are able to respond to more than one of the basic taste stimuli, each type responds most strongly to one of them — SWEET, SOUR, SALT or BITTER.

Substances in solution - e.g. in SALIVA or in serous secretion of Von Ebner's glands — enter pore → stimulate receptors for taste (Gustatory cells) → Nerve impulses relayed in nerves of taste → Taste Centres in Brain

Other tastes are probably due to combinations of these with smell or with ordinary skin sensations.

PATHWAYS and CENTRES for TASTE

The RECEPTORS for TASTE are linked by a chain of 3 Neurones with the RECEIVING CENTRES for TASTE in the CEREBRAL CORTEX.

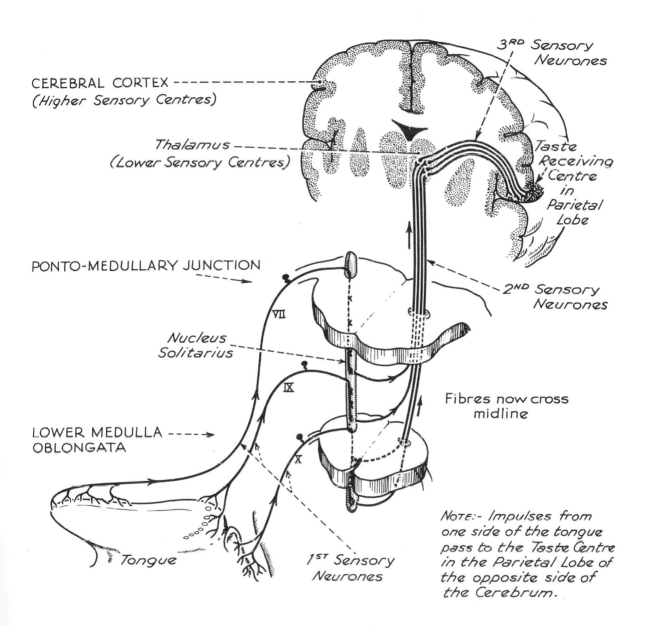

CEREBRAL CORTEX
(Higher Sensory Centres)

3RD Sensory Neurones

Thalamus
(Lower Sensory Centres)

Taste Receiving Centre in Parietal Lobe

PONTO-MEDULLARY JUNCTION

2ND Sensory Neurones

VII

Nucleus Solitarius

IX

Fibres now cross midline

LOWER MEDULLA OBLONGATA

X

Tongue

1ST Sensory Neurones

NOTE:- Impulses from one side of the tongue pass to the Taste Centre in the Parietal Lobe of the opposite side of the Cerebrum.

EYE

STRUCTURE   *FUNCTION of PARTS*

The Eyeball has three coats :-

1. OUTER or SCLEROTIC COAT   **PROTECTIVE LAYER**

Tough fibrous tissue of "White of Eye". ···· *Preserves shape of eyeball and protects delicate inner layers.*

Transparent CORNEA in front. ···· *Allows passage of LIGHT RAYS.*

[Extrinsic muscles are attached to sclera. ···· *Permit and limit movements of eyeball within ORBIT.*]

2. MIDDLE or VASCULAR PIGMENTED COAT   **LAYER of SUPPLY**

Contains main arteries and veins of eyeball.

Circular opening at front—PUPIL.

Coloured muscular ring—IRIS—surrounds pupil. ···· *Controls size of pupil and amount of light entering eye.*

CILIARY BODY. ···· *Produces AQUEOUS HUMOUR.*

CILIARY MUSCLE. ···· *Contracts and moves forwards.*

SUSPENSORY LIGAMENT suspends CRYSTALLINE LENS. ···· *Relaxes to allow curvature of lens to alter for accommodation for NEAR VISION.*

CHOROID—Posterior 5/6 of Vascular Coat. ···· *Brings light rays to focus on light-sensitive RETINA.*

3. INNER or NERVOUS COAT—the RETINA   **LIGHT-SENSITIVE LAYER**

Lines back of eye. Contains RECEPTORS for VISION ···· *Highly specialized to respond to stimulation by light. Convert light energy into nerve impulses.* ⟶

OPTIC NERVE
Conveys these impulses to VISUAL CENTRES in OCCIPITAL (posterior) part of BRAIN.

CORNEA

ANTERIOR CHAMBER (AQUEOUS HUMOUR)

IRIS

SUSPENSORY LIGAMENT

CILIARY BODY

LENS

(VITREOUS HUMOUR)

RETINA
CHOROID
SCLERA

Main blood vessels to retina enter and leave here.

CENTRAL ARTERY of the RETINA

PROTECTION of the EYE

The hidden posterior ⁴⁄₅ of the eyeball is encased in a bony socket – the ORBITAL CAVITY.   A thick layer of *Areolar* and *Adipose* tissue forms a cushion between bone and eyeball. The exposed anterior ⅕ of the eyeball is protected from injury by:-

The <u>EYELIDS</u> ------------------------ close reflexly to protect eye from dust
 Fringed with E*YELASHES*   *and other foreign particles.*

<u>CONJUNCTIVA</u> ------------------- *smooth surfaces which glide over each*
 A delicate membrane lining eyelids   *other when lids open and close.*
and covering exposed surface of eye.

<u>LACRIMAL GLANDS</u> --------------- *continuously secrete TEARS. These*
              flow over, wash and lubricate surface
              of eye. They contain an ENZYME –
              LYSOZYME – which destroys bacteria.

<u>TARSAL GLANDS</u> ---------------------- *secrete a fluid to prevent lids*
              from sticking together

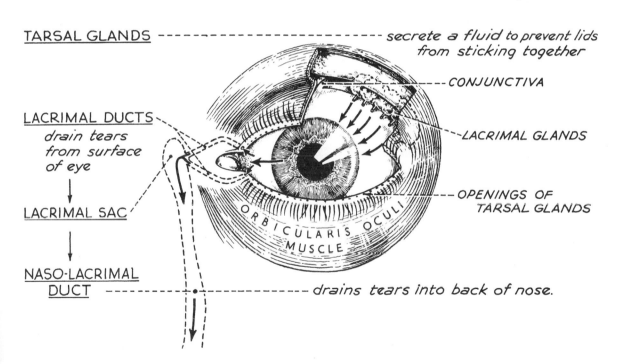

CONJUNCTIVA

LACRIMAL GLANDS

<u>LACRIMAL DUCTS</u>
 drain tears
 from surface
 of eye

OPENINGS OF
TARSAL GLANDS

ORBICULARIS OCULI
MUSCLE

<u>LACRIMAL SAC</u>

<u>NASO-LACRIMAL</u>
<u>DUCT</u> ------------------ *drains tears into back of nose.*

MUSCLES of EYE

The EYEBALLS are moved by SMALL MUSCLES which link the SCLEROTIC COAT to the BONY SOCKET.

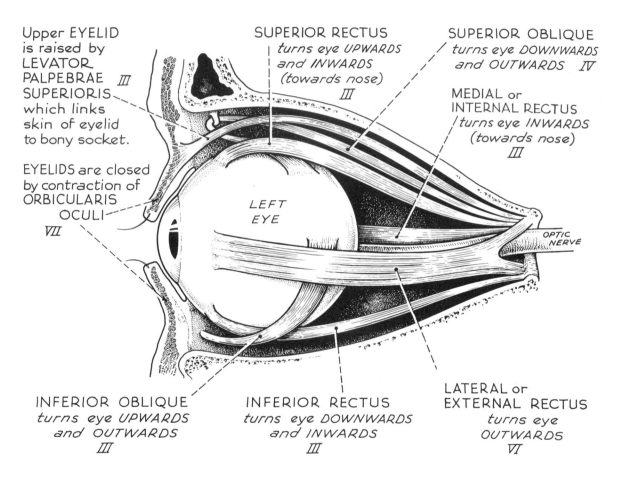

Upper EYELID is raised by LEVATOR PALPEBRAE SUPERIORIS *III* which links skin of eyelid to bony socket.

EYELIDS are closed by contraction of ORBICULARIS OCULI *VII*

SUPERIOR RECTUS
turns eye UPWARDS and INWARDS (towards nose)
III

SUPERIOR OBLIQUE
turns eye DOWNWARDS and OUTWARDS   IV

MEDIAL or INTERNAL RECTUS
turns eye INWARDS (towards nose)
III

LEFT EYE

OPTIC NERVE

INFERIOR OBLIQUE
turns eye UPWARDS and OUTWARDS
III

INFERIOR RECTUS
turns eye DOWNWARDS and INWARDS
III

LATERAL or EXTERNAL RECTUS
turns eye OUTWARDS
VI

Acting together, the Extrinsic muscles of the Eyeballs can bring about ROTATORY movements of the Eyes.
The Extrinsic muscles are supplied by motor fibres from Cranial Nerves III, IV and VI.

CONTROL of EYE MOVEMENTS

Both eyes normally move together so that images continue to fall on corresponding points of both retinae.

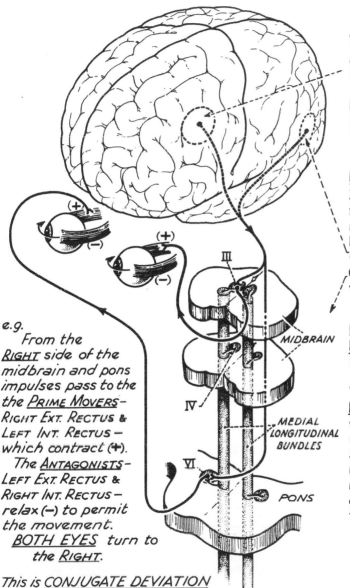

VOLUNTARY EYE MOVEMENTS
are initiated in motor centres in FRONTAL LOBES.

Impulses from <u>one side</u> of the Cerebral Cortex turn <u>both eyes</u> to the <u>other side</u> of Visual Field.

REFLEX EYE MOVEMENTS
Two Groups – (1) Those in response to Visual Stimuli. (2) Those in response to Non-Visual Stimuli. In control of these are :-
CENTRES in OCCIPITAL LOBES:
CENTRES in MIDBRAIN and PONS which give rise to CRANIAL NERVES III, IV and VI.

Impulses from <u>one side</u> of the <u>Midbrain and Pons</u> turn eyes to the <u>same side</u>.

These centres are <u>closely linked</u> with each other and with <u>HIGHER and LOWER</u> centres in the Central Nervous System, so that the eyes are moved reflexly in response to many stimuli, e.g. loud noises or proprioceptive messages from vestibular organs.

e.g.
From the <u>RIGHT</u> side of the midbrain and pons impulses pass to the the <u>PRIME MOVERS</u>- RIGHT EXT. RECTUS & LEFT INT. RECTUS – which contract (+).
The <u>ANTAGONISTS</u>- LEFT EXT. RECTUS & RIGHT INT. RECTUS – relax (−) to permit the movement.
<u>BOTH EYES</u> turn to the <u>RIGHT</u>.

This is <u>CONJUGATE DEVIATION</u>

MIDBRAIN

MEDIAL LONGITUDINAL BUNDLES

PONS

III

IV

VI

227

IRIS, LENS and CILIARY BODY

The __IRIS__ is a muscular diaphragm with a central opening - the PUPIL.

The __LENS__ is a transparent biconvex crystalline disc.

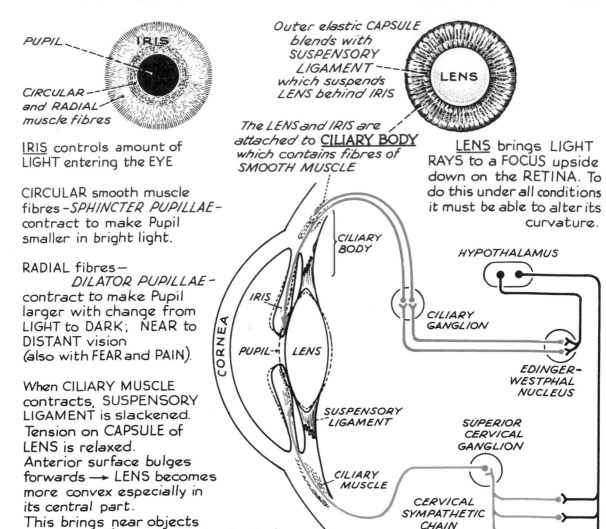

PUPIL

IRIS

CIRCULAR and RADIAL muscle fibres

Outer elastic CAPSULE blends with SUSPENSORY LIGAMENT which suspends LENS behind IRIS

LENS

The LENS and IRIS are attached to __CILIARY BODY__ which contains fibres of SMOOTH MUSCLE

__IRIS__ controls amount of LIGHT entering the EYE

CIRCULAR smooth muscle fibres –*SPHINCTER PUPILLAE*– contract to make Pupil smaller in bright light.

RADIAL fibres—
        DILATOR PUPILLAE–
contract to make Pupil larger with change from LIGHT to DARK; NEAR to DISTANT vision
(also with FEAR and PAIN).

When CILIARY MUSCLE contracts, SUSPENSORY LIGAMENT is slackened. Tension on CAPSULE of LENS is relaxed. Anterior surface bulges forwards ⟶ LENS becomes more convex especially in its central part.
This brings near objects into focus. (Accommodation Reflex)

LENS brings LIGHT RAYS to a FOCUS upside down on the RETINA. To do this under all conditions it must be able to alter its curvature.

CILIARY BODY

HYPOTHALAMUS

IRIS

CILIARY GANGLION

CORNEA

PUPIL

LENS

EDINGER-WESTPHAL NUCLEUS

SUPERIOR CERVICAL GANGLION

SUSPENSORY LIGAMENT

CILIARY MUSCLE

CERVICAL SYMPATHETIC CHAIN

These changes are brought about reflexly. The ingoing impulses travel in the Optic nerves. The outgoing motor impulses travel in Parasympathetic to Ciliary Body and Sphincter Pupillae and in Sympathetic to Dilator Pupillae.

ACTION of LENS

The normal lens brings light rays to a sharp focus upside down on the retina. It can do this whether we are looking at an object far away or one close at hand. The curvature increases reflexly to accommodate for near vision.

Rays of light coming from every point of a DISTANT object (over 20 feet away) are PARALLEL.

They pass through the CORNEA, AQUEOUS HUMOUR

and LENS which refract them to a sharp focus — upside down and reversed from side to side – on the Retina.

The conscious mind learns to interpret the image and project it to its true position in space.

Rays of light coming from a NEAR object (less than 20 feet away) RADIATE from every point

A more convex lens is required to bring these rays to a sharp focus on the Retina.

If the EYEBALL is <u>too short</u>, rays from a distant object are brought into focus BEHIND the Retina when the ciliary muscle is relaxed.

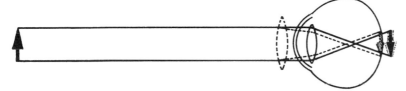

Vision is blurred.

This is longsightedness or HYPERMETROPIA.

The longsighted eye has to accommodate even for distant vision: i.e. Ciliary muscles contract to give a more convex lens and distant objects are then seen clearly. This limits amount of accommodating power left for near objects and the nearest point for sharp vision is then further away. It can be corrected by fitting spectacles with an additional convex lens.

If the EYEBALL is <u>too long</u>, rays from a distant object are brought into focus IN FRONT of the Retina.

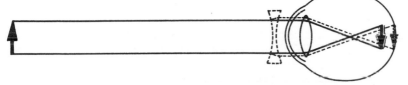

This is shortsightedness or MYOPIA — only objects near the eye can be seen clearly.

It can be corrected by using a concave lens.

FUNDUS OCULI

Part of the RETINA can be seen by means of an instrument – the OPHTHALMOSCOPE – which shines a beam of light through the PUPIL of the eye on to the RETINA.

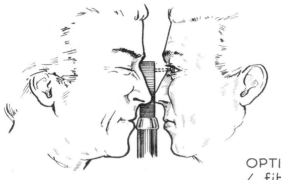

The part of the Retina seen in this way is called the <u>FUNDUS OCULI</u>.

OPTIC PAPILLA – or 'BLIND SPOT' – The nerve fibres from all parts of the retina converge on this area to leave the eyeball as the OPTIC NERVE. It has no RODS or CONES and therefore is not itself sensitive to light.

RETINAL BLOOD VESSELS enter or leave the eyeball here.

MACULA LUTEA – or 'YELLOW SPOT' – with

FOVEA CENTRALIS – area of acute vision – contains CONES only (the Receptors stimulated in BRIGHT and COLOURED LIGHT). When we look at an object the eyes are directed so that the image will fall on the fovea of each eye.

EXTRAFOVEAL part of Retina – area of less acute vision – CONES become fewer – RODS (the Receptors stimulated in DIM LIGHT with no discrimination between COLOURS) become more numerous towards the peripheral part of the Retina.

LEFT FUNDUS

230

RETINA

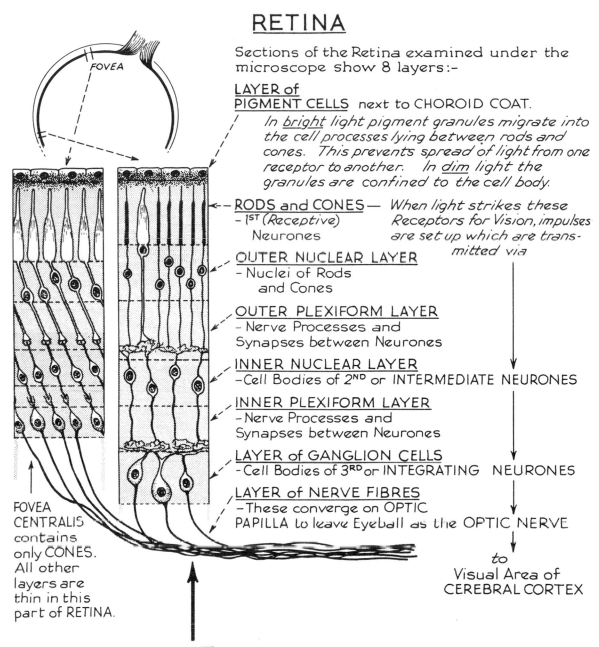

Sections of the Retina examined under the
microscope show 8 layers :-

<u>LAYER of
PIGMENT CELLS</u> next to CHOROID COAT.

*In <u>bright</u> light pigment granules migrate into
the cell processes lying between rods and
cones. This prevents spread of light from one
receptor to another. In <u>dim</u> light the
granules are confined to the cell body.*

<u>RODS and CONES</u> — *When light strikes these*
- 1ST (Receptive)   *Receptors for Vision, impulses*
  Neurones   *are set up which are trans-*
                       mitted via

<u>OUTER NUCLEAR LAYER</u>
- Nuclei of Rods
  and Cones

<u>OUTER PLEXIFORM LAYER</u>
- Nerve Processes and
Synapses between Neurones

<u>INNER NUCLEAR LAYER</u>
-Cell Bodies of 2ND or INTERMEDIATE NEURONES

<u>INNER PLEXIFORM LAYER</u>
- Nerve Processes and
Synapses between Neurones

<u>LAYER of GANGLION CELLS</u>
-Cell Bodies of 3RD or INTEGRATING NEURONES

<u>LAYER of NERVE FIBRES</u>
-These converge on OPTIC
PAPILLA to leave Eyeball as the OPTIC NERVE

to
Visual Area of
CEREBRAL CORTEX

FOVEA
CENTRALIS
contains
only CONES.
All other
layers are
thin in this
part of RETINA.

Note:- LIGHT rays must pass through all
these layers except the pigment
cell layer to reach and stimulate RECEPTORS.

MECHANISM of VISION

WHITE LIGHT is really due to the fusion of *coloured lights*. These coloured lights are separated by shining a beam of white light through a glass prism. This is called the VISIBLE SPECTRUM.

7,800Å	7,000Å			6,000Å				5,000Å			3,900Å
RED		ORANGE	YELLOW	YELL.-GR.	GREEN	BLUE-GREEN	BLUE	VIOLET			

If light is bright or intense the spectrum appears brightest to man's eye in the orange band (6,100 Å).

If brightness or intensity of light source is gradually reduced, colour perception is gradually lost and the spectrum appears as a luminous band with a very dark area in the red band and with its brightest part in the green band (5,300Å).

Visual Receptors contain PIGMENTS which break down chemically in the presence of specific wavelengths of light. This forms a chemical stimulus which in some way triggers off nerve impulses which travel from the retina to the cerebral cortex.

SCOTOPIC VISION is vision in DIM LIGHT.   It depends on the RODS

RODS are of one type ——————and give——————MONOCHROMATIC VISION

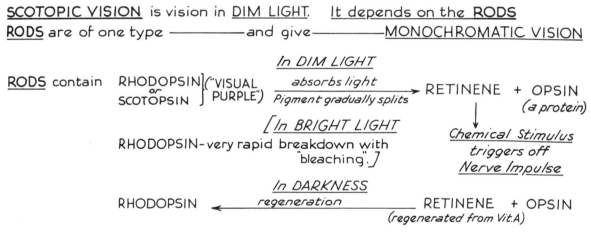

When a person passes from a brightly lit scene to darkness he is temporarily blinded. After about ½ hour he sees well. This adjustment or increase in sensitivity is called DARK ADAPTATION.

As light brightness or intensity increases rods lose their sensitivity and cease to respond.

232

MECHANISM of VISION

PHOTOPIC VISION is vision in BRIGHT LIGHT. It depends on the CONES

CONES are thought to be of 3 types —— giving TRICHROMATIC VISION.
      Each type with a
        different
     photosensitive
       VISUAL
       PIGMENT with its own wavelength to which it is sensitive, which
                it absorbs and by which it is broken down to form the
                                chemical stimulus.

Note:- The Photosensitive pigments in cones have not yet been isolated.

"RED" receptors absorb YELLOW—ORANGE light
"GREEN" receptors absorb   GREEN   light
"BLUE" receptors absorb   BLUE   light

All 3 types of "PHOTOPSINS" are probably stimulated in roughly equal
                  proportions when WHITE light falls on retina:
2 or more in varying proportions when OTHER COLOURS of light fall on
                                     retina.

The various types of colour blindness could be explained in terms of the absence or deficiency of one or more of these special receptors.

[*A colour sensation has 3 qualities:-*
    HUE ——————— depends largely on wavelength.
    SATURATION — purity —
           A "saturated" colour has no white light mixed with it.
           An "unsaturated" colour has some white light mixed with it.
    INTENSITY — brightness — depends largely on "strength" of the light.]

As the intensity of light is reduced the Cones cease to respond and the Rods take over.

When a person passes from darkness to bright light he is dazzled but after a short time he sees well again.
   This adjustment or decrease in sensitivity on exposure to bright light is called LIGHT ADAPTATION.

VISUAL PATHWAYS to the BRAIN

The RECEPTORS for VISION are linked by a chain of Neurones with RECEIVING and INTEGRATING CENTRES in the OCCIPITAL LOBES of the CEREBRAL CORTEX.

The two VISUAL FIELDS are slightly DISSIMILAR but OVERLAP. ------------>

LIGHT RAYS from LEFT side of Visual fields fall on the INNER (Nasal) HALF of LEFT RETINA
                    and
OUTER (Temporal) HALF of RIGHT RETINA.

Impulses are relayed from RECEPTOR Neurones by INTERMEDIATE to INTEGRATING Neurones whose axons form the OPTIC NERVES. ------------->

Fibres from the NASAL HALF of each RETINA cross in OPTIC CHIASMA.------

Fibres from the TEMPORAL HALF of each RETINA remain uncrossed.

Most fibres SYNAPSE with Neurones in the LOWER VISUAL CENTRES – LATERAL GENICULATE BODY ------------

[Some fibres synapse in Pretectal Region, some synapse in Superior Colliculus — these form afferents for the OPTIC REFLEXES.]

The axons of these Neurones form OPTIC RADIATIONS
They relay VISUAL IMPULSES to HIGHER VISUAL CENTRES in OCCIPITAL LOBES where the images from both eyes are integrated into a single SENSATION.

MIDBRAIN

Note:- One side of the OCCIPITAL CORTEX receives impressions from the FIELD of VISION on the opposite side.

234

STEREOSCOPIC VISION

When we look at some object or scene the view seen by the RIGHT EYE is slightly different from the view seen by the LEFT EYE.

These two DISSIMILAR RETINAL IMAGES are fused in the Visual Centres of the Brain to give a 3-dimensional picture – an appreciation of DEPTH as well as of HEIGHT and WIDTH.

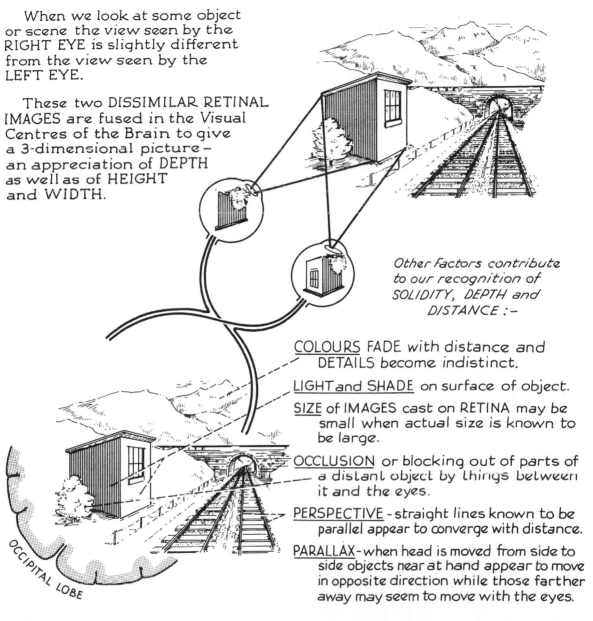

Other Factors contribute to our recognition of SOLIDITY, DEPTH and DISTANCE :–

COLOURS FADE with distance and DETAILS become indistinct.

LIGHT and SHADE on surface of object.

SIZE of IMAGES cast on RETINA may be small when actual size is known to be large.

OCCLUSION or blocking out of parts of a distant object by things between it and the eyes.

PERSPECTIVE - straight lines known to be parallel appear to converge with distance.

PARALLAX - when head is moved from side to side objects near at hand appear to move in opposite direction while those farther away may seem to move with the eyes.

OCCIPITAL LOBE

By complex mental processes these points are interpreted in terms of Distance and Depth.

LIGHT REFLEX

When LIGHT falls on the RETINA the PUPILS CONTRACT.

Impulses from
the Receptors in
the Retina pass
in the
OPTIC NERVE
to the
PRETECTAL
NUCLEUS
in the midbrain
of the same side

Impulses then
pass to the
EDINGER-WESTPHAL
NUCLEUS
of the 3RD Cranial
Nerves of both sides

Impulses now travel
in efferent
Parasympathetic motor
nerves to cause
contraction of both
SPHINCTER
PUPILLAE

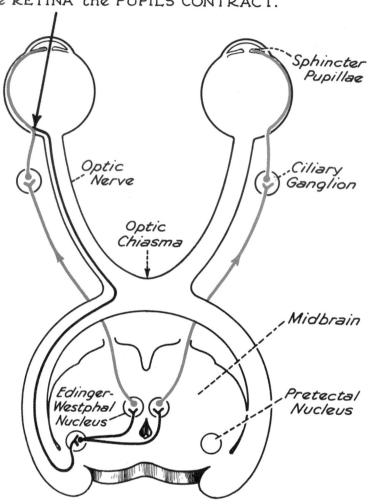

DIRECT LIGHT REFLEX

When light shines into
one eye (as in diagram)
the PUPIL of that eye contracts ——— the PUPIL of the other eye also
contracts.

CONSENSUAL LIGHT REFLEX

This cuts down the amount of light entering the eyes and protects
the Retinae from excessive stimulation. It also increases depth of
focus and improves the sharpness of the Retinal images.

EAR

The ear has 3 separate parts, each with different rôles in the mechanism of HEARING:-

---- **OUTER EAR** ---

MIDDLE EAR

Small chamber deep within the *TEMPORAL BONE*.
Contains 3 small bones – the *AUDITORY OSSICLES* - connected to form a small *LEVER*.
MALLEUS attached to drum and to *INCUS* linked to *STAPES* which fits into *OVAL WINDOW*.

---- **INNER EAR** ---

Contains *ORGANS* of *EQUILIBRIUM* and *HEARING*.

SEMICIRCULAR CANALS (*non-auditory part of inner ear — concerned with EQUILIBRIUM sense*).

COCHLEA (in Spiral Canal) contains *RECEPTORS* for *HEARING (Organ of Corti)*.

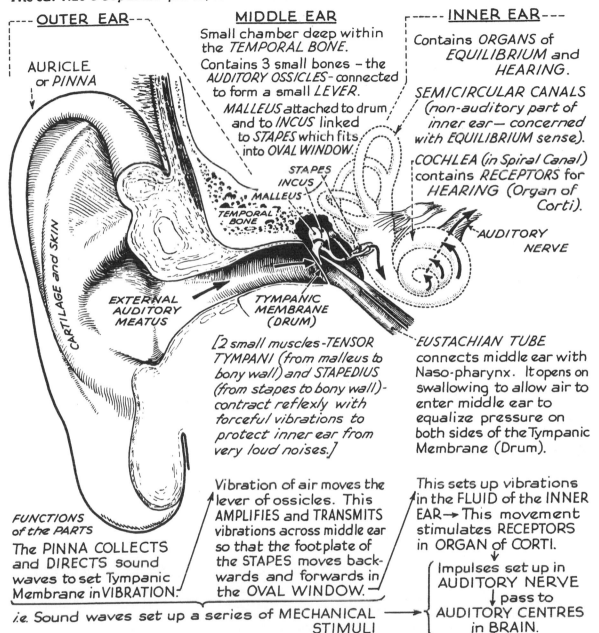

AURICLE or *PINNA*

CARTILAGE and SKIN

STAPES
INCUS
MALLEUS
TEMPORAL BONE

EXTERNAL AUDITORY MEATUS

TYMPANIC MEMBRANE (DRUM)

AUDITORY NERVE

[2 small muscles -TENSOR TYMPANI (from malleus to bony wall) and STAPEDIUS (from stapes to bony wall)- contract reflexly with forceful vibrations to protect inner ear from very loud noises.]

EUSTACHIAN TUBE connects middle ear with Naso-pharynx. It opens on swallowing to allow air to enter middle ear to equalize pressure on both sides of the Tympanic Membrane (Drum).

FUNCTIONS of the PARTS
The PINNA COLLECTS and DIRECTS sound waves to set Tympanic Membrane in VIBRATION.

Vibration of air moves the lever of ossicles. This AMPLIFIES and TRANSMITS vibrations across middle ear so that the footplate of the STAPES moves backwards and forwards in the OVAL WINDOW.

This sets up vibrations in the FLUID of the INNER EAR→ This movement stimulates RECEPTORS in ORGAN of CORTI.

Impulses set up in AUDITORY NERVE ↓ pass to AUDITORY CENTRES in BRAIN.

i.e. Sound waves set up a series of MECHANICAL STIMULI →

COCHLEA

The Cochlea is the essential organ of HEARING.

It consists of :–

The BONY COCHLEA which spirals 2¾ times round central pillar of bone.

The MEMBRANOUS COCHLEA which is enclosed between the VESTIBULAR and BASILAR Membranes

These spiral compartments are filled with FLUID

STRIA VASCULARIS (pigmented, granular cells with profuse blood supply) secretes ENDOLYMPH of SCALA MEDIA

SCALA VESTIBULI and SCALA TYMPANI contain PERILYMPH

COCHLEAR NERVE

APEX

BASE

The BASILAR MEMBRANE is broad at the Apex of the Cochlea and short at the Base.

The EXTERNAL SPIRAL LIGAMENT attaches the Basilar Membrane to the bony wall. It is more powerful at the Base of Cochlea than at the Apex.

On the Basilar membrane lies the ORGAN of CORTI. This contains the RECEPTORS for HEARING.

The tips of their hair-like processes are embedded in the TECTORIAL MEMBRANE which floats in Endolymph

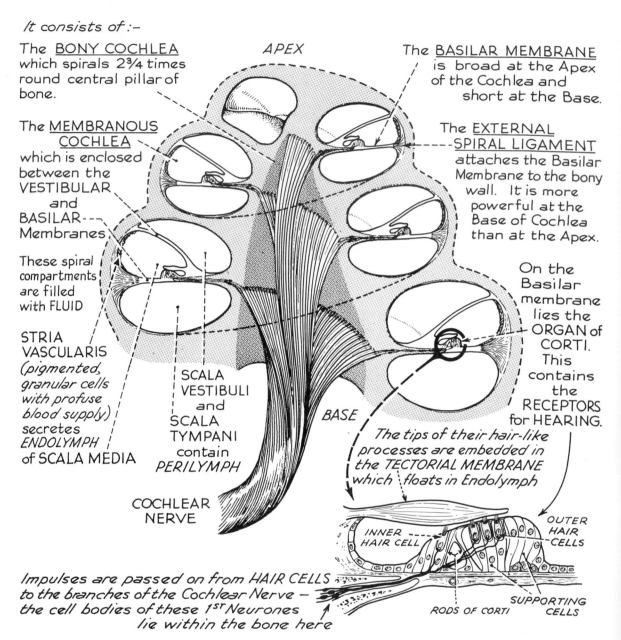

OUTER HAIR CELLS

INNER HAIR CELL

RODS OF CORTI

SUPPORTING CELLS

Impulses are passed on from HAIR CELLS to the branches of the Cochlear Nerve – the cell bodies of these 1ST Neurones lie within the bone here

238

MECHANISM of HEARING

This is most readily understood if the COCHLEA is imagined as straightened out:—

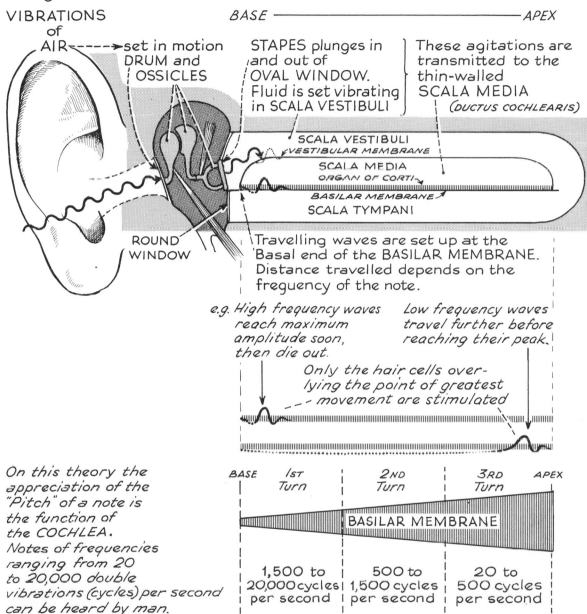

VIBRATIONS of AIR ---→ set in motion DRUM and OSSICLES

BASE ——————————— APEX

STAPES plunges in and out of OVAL WINDOW. Fluid is set vibrating in SCALA VESTIBULI

These agitations are transmitted to the thin-walled SCALA MEDIA *(DUCTUS COCHLEARIS)*

SCALA VESTIBULI
VESTIBULAR MEMBRANE
SCALA MEDIA
ORGAN OF CORTI
BASILAR MEMBRANE
SCALA TYMPANI

ROUND WINDOW

Travelling waves are set up at the Basal end of the BASILAR MEMBRANE. Distance travelled depends on the frequency of the note.

e.g. High frequency waves reach maximum amplitude soon, then die out.

Low frequency waves travel further before reaching their peak.

Only the hair cells over-lying the point of greatest movement are stimulated

On this theory the appreciation of the "Pitch" of a note is the function of the COCHLEA. Notes of frequencies ranging from 20 to 20,000 double vibrations (cycles) per second can be heard by man.

BASE   1ST Turn   2ND Turn   3RD Turn   APEX

BASILAR MEMBRANE

| 1,500 to 20,000 cycles per second | 500 to 1,500 cycles per second | 20 to 500 cycles per second |

239

AUDITORY PATHWAYS to BRAIN

The RECEPTORS for HEARING are linked by a chain of NEURONES with the RECEIVING CENTRES for HEARING in the TEMPORAL LOBES of the CEREBRAL CORTEX.

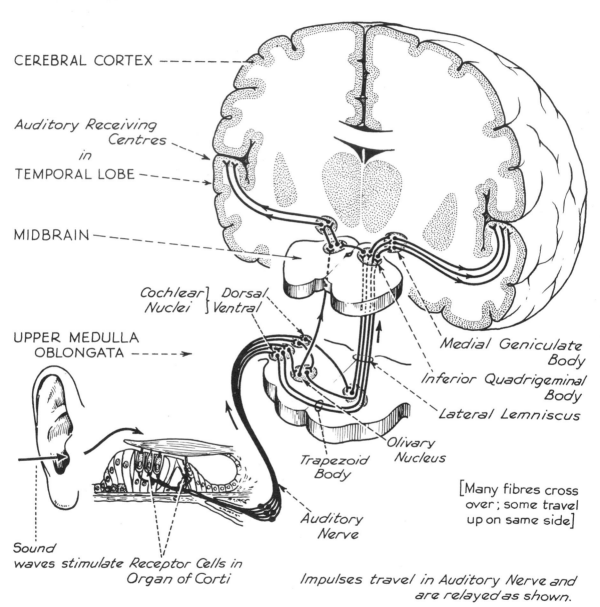

CEREBRAL CORTEX - - - - -

Auditory Receiving
Centres
in
TEMPORAL LOBE - - -

MIDBRAIN - - -

Cochlear] Dorsal
Nuclei]Ventral

UPPER MEDULLA
OBLONGATA - - - - ->

Medial Geniculate
Body

Inferior Quadrigeminal
Body

Lateral Lemniscus

Olivary
Nucleus

Trapezoid
Body

Auditory
Nerve

[Many fibres cross
over; some travel
up on same side]

Sound
waves stimulate Receptor Cells in
Organ of Corti

Impulses travel in Auditory Nerve and
are relayed as shown.

240

SPECIAL PROPRIOCEPTORS

The "Special" Proprioceptors of the body are found in the Non-Auditory part of the Inner Ear — the Labyrinth. They are stimulated by movements or change of position of the head in space.

Three **SEMICIRCULAR CANALS** — *in each inner ear – one in each of three planes of space* – contain Receptors.

Membranous tubes containing Endolymph

These receptors, situated in the Ampulla of each canal, are stimulated mechanically by the *starting or stopping of rotatory movements* of the head in space.

One end of each canal has a swelling - the AMPULLA

Both ends of each canal open into the UTRICLE

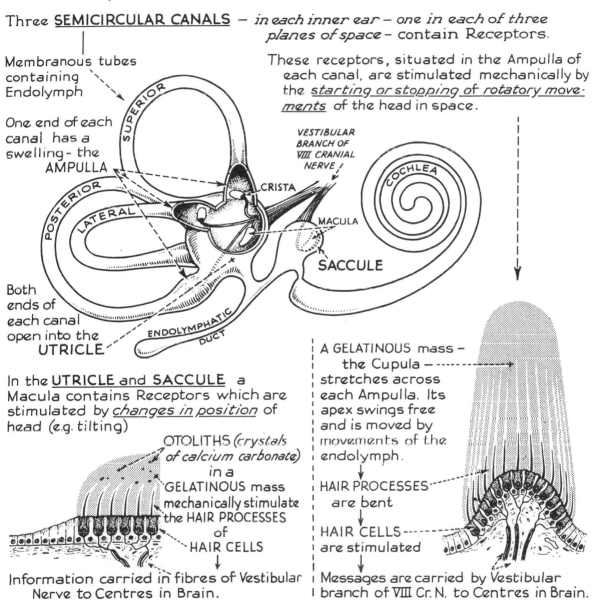

In the **UTRICLE** and **SACCULE** a Macula contains Receptors which are stimulated by *changes in position* of head (e.g. tilting)

OTOLITHS *(crystals of calcium carbonate)* in a GELATINOUS mass mechanically stimulate the HAIR PROCESSES of HAIR CELLS

Information carried in fibres of Vestibular Nerve to Centres in Brain.

A GELATINOUS mass — the Cupula — stretches across each Ampulla. Its apex swings free and is moved by movements of the endolymph.

HAIR PROCESSES are bent

HAIR CELLS are stimulated

Messages are carried by Vestibular branch of VIII Cr. N. to Centres in Brain.

ORGAN of EQUILIBRIUM: MECHANISM of ACTION

At Rest

RIGHT
Lateral
Horizontal
Canal

-------- AMPULLA --------
------- CUPULA -------
------- CRISTA -------
BRANCHES --
of VESTIBULAR
NERVE

LEFT
Lateral
Horizontal
Canal

When head starts to rotate (e.g. to the right) the endolymph in the semicircular canals which lie at right angles to the axis of rotation tends to lag behind the movement of the head — and the

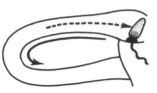

CUPULA is displaced,
HAIR CELLS are stimulated and
INGOING impulses form afferent pathways
for Reflexes leading to alterations in tone
of muscles in neck, trunk and limbs to
avoid body losing balance.

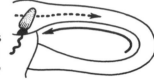

After the initial inertia is overcome the endolymph no longer lags behind the movement of the head — and the

CUPULA is no longer displaced.
HAIR CELLS are no longer bent and
stimulated.
Nerve fibres no longer send signals
to medulla and cerebellum.

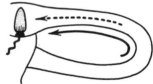

When head stops rotating the endolymph tends to continue to rotate and the

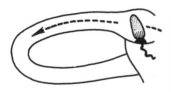

CUPULA is displaced in the opposite
direction → HAIR CELLS are bent →
Nerve fibres signal rotation of head
to left → Individual feels for a
moment as though he is rotating
in opposite direction — when in fact
he has ceased to rotate.

   Rotating the head round a horizontal axis — i.e. tilting it backwards and forwards such as happens in the pitch and roll of a ship — stimulates the vertical canals. This can lead to "motion sickness".
   Stimulation of the Semicircular canals also causes movements of the eyes to avoid too much displacement of the image being cast on the retinae.
   During rotation there is a slow movement of the eyes in the direction opposite to that of rotation, then a quick return to the normal position. This is NYSTAGMUS. It continues for a short time after movement has ceased.

242

VESTIBULAR PATHWAYS to BRAIN

The Special Proprioceptive end-organs in the LABYRINTH are linked through the VESTIBULAR NUCLEI with RECEIVING and INTEGRATING CENTRES in the CEREBELLUM, and with MOTOR CENTRES in the MIDBRAIN and SPINAL CORD through which they initiate reflex muscular movements of EYES, HEAD and NECK, TRUNK and LIMB muscles to adjust balance and posture.

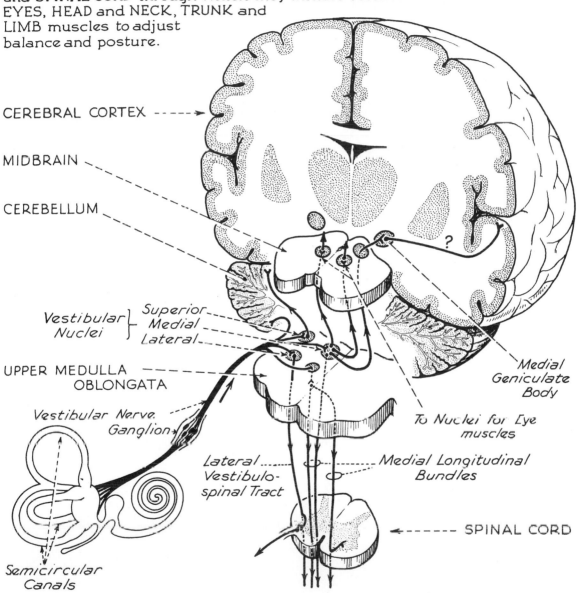

CEREBRAL CORTEX - - - - >

MIDBRAIN

CEREBELLUM

Vestibular ⎱ Superior
Nuclei ⎰ Medial
         Lateral

UPPER MEDULLA OBLONGATA

Vestibular Nerve Ganglion

Semicircular Canals

Lateral Vestibulo-spinal Tract

Medial Geniculate Body

To Nuclei for Eye muscles

Medial Longitudinal Bundles

SPINAL CORD

?

GENERAL PROPRIOCEPTORS

Proprioceptors are the sense organs stimulated by MOVEMENT of the body itself. They make us aware of the movement or position of the body in space and of the various parts of the body to each other. They are important as ingoing afferent pathways in reflexes for adjusting Posture and Tone.

General Proprioceptors are found in SKELETAL MUSCLES, TENDONS and JOINTS.

GOLGI ORGAN — in tendons —
stimulated by tension which occurs when muscle is STRETCHED and when it is CONTRACTED

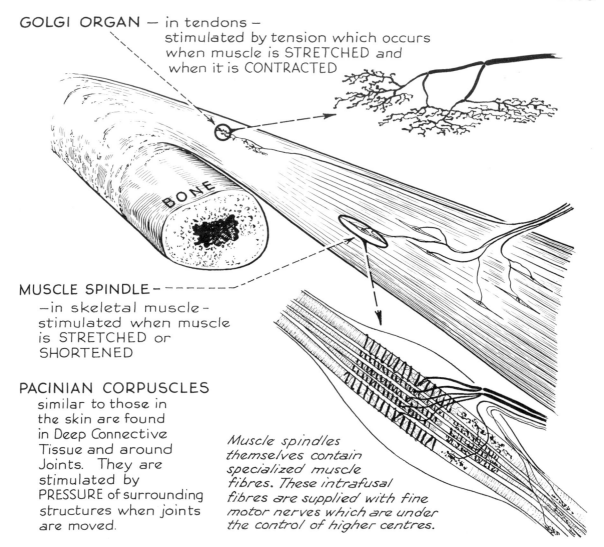

MUSCLE SPINDLE — — — — — —
—in skeletal muscle— stimulated when muscle is STRETCHED or SHORTENED

PACINIAN CORPUSCLES
similar to those in the skin are found in Deep Connective Tissue and around Joints. They are stimulated by PRESSURE of surrounding structures when joints are moved.

Muscle spindles themselves contain specialized muscle fibres. These intrafusal fibres are supplied with fine motor nerves which are under the control of higher centres.

244

PROPRIOCEPTOR PATHWAYS to BRAIN

General Proprioceptor end-organs may be linked with centres in
(a) CEREBELLUM or (b) PARIETAL LOBE of CEREBRAL CORTEX by a chain
of 3 Neurones.

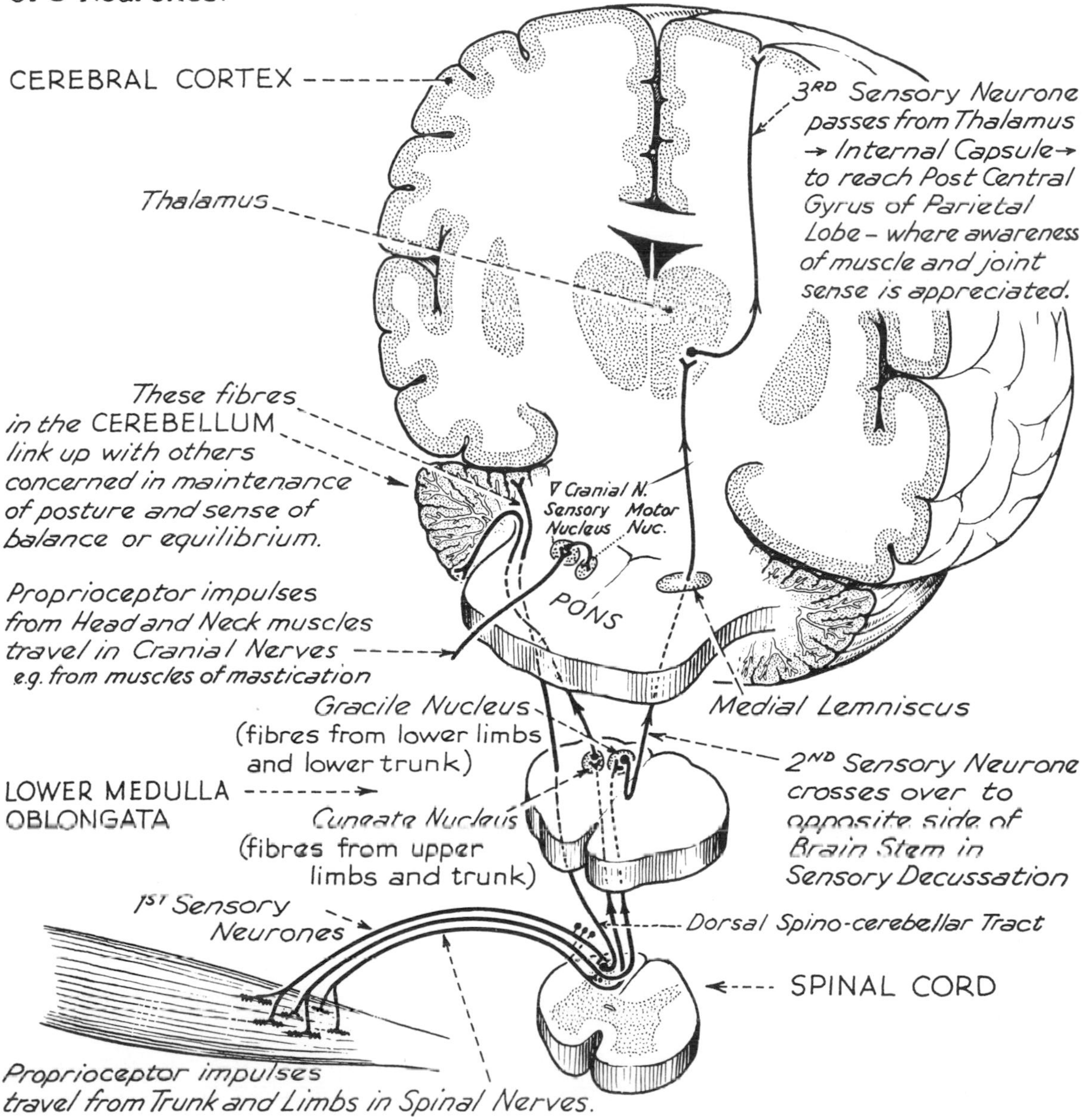

CEREBRAL CORTEX - - - - - - -

*3ᴿᴰ Sensory Neurone
passes from Thalamus
→ Internal Capsule →
to reach Post Central
Gyrus of Parietal
Lobe – where awareness
of muscle and joint
sense is appreciated.*

Thalamus - - - -

*These fibres
in the CEREBELLUM
link up with others
concerned in maintenance
of posture and sense of
balance or equilibrium.*

V Cranial N.
Sensory Motor
Nucleus Nuc.

PONS

*Proprioceptor impulses
from Head and Neck muscles
travel in Cranial Nerves - - - -
e.g. from muscles of mastication*

Gracile Nucleus
*(fibres from lower limbs
and lower trunk)*

Medial Lemniscus

LOWER MEDULLA - - - - - - - →
OBLONGATA

Cuneate Nucleus
*(fibres from upper
limbs and trunk)*

*2ᴺᴰ Sensory Neurone
crosses over to
opposite side of
Brain Stem in
Sensory Decussation*

*1ˢᵀ Sensory
Neurones*

Dorsal Spino-cerebellar Tract

- - - - SPINAL CORD

*Proprioceptor impulses
travel from Trunk and Limbs in Spinal Nerves.*

245

CUTANEOUS SENSATION

There are 5 basic skin sensations — TOUCH, PRESSURE, PAIN, WARMTH and COLD. There is much controversy as to how these are registered. In some areas they appear to be served by special nerve endings (sensory receptors or end-organs) in the SKIN. These end-organs are not uniformly distributed over the whole body surface. *(Eg. Touch "endings" are very numerous in hands and feet but are much less frequent in the skin of the back.)*

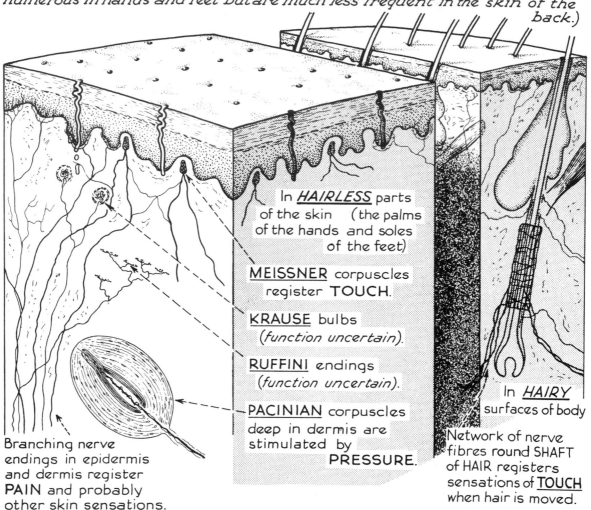

In *HAIRLESS* parts of the skin (the palms of the hands and soles of the feet)

MEISSNER corpuscles register TOUCH.

KRAUSE bulbs (function uncertain).

RUFFINI endings (function uncertain).

PACINIAN corpuscles deep in dermis are stimulated by PRESSURE.

Branching nerve endings in epidermis and dermis register PAIN and probably other skin sensations.

In *HAIRY* surfaces of body

Network of nerve fibres round SHAFT of HAIR registers sensations of TOUCH when hair is moved.

Tickling, itching, softness, hardness, wetness are probably due to stimulation of two or more of these special endings and to a blending of the sensations in the Brain.

246

SENSORY PATHWAYS from SKIN of FACE

The Receptors or nerve endings for ORDINARY SKIN SENSATIONS are linked by 3 Neurones with Receiving Centres in the PARIETAL LOBES.

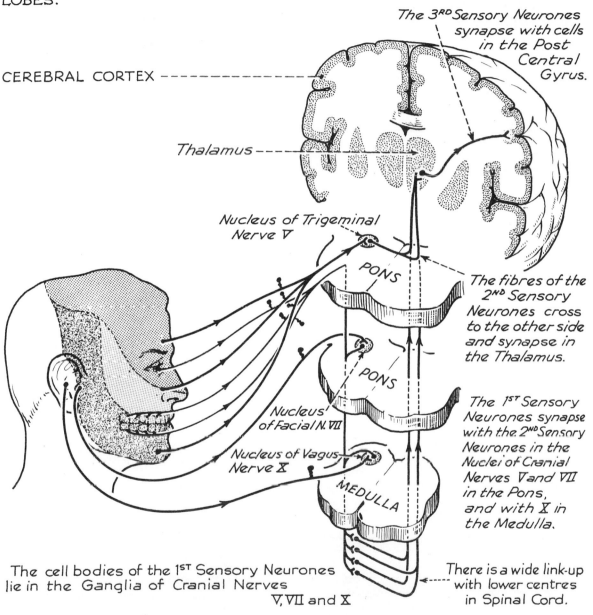

The 3ᴿᴰ Sensory Neurones synapse with cells in the Post Central Gyrus.

CEREBRAL CORTEX -----------

Thalamus ------

Nucleus of Trigeminal Nerve V

PONS

The fibres of the 2ᴺᴰ Sensory Neurones cross to the other side and synapse in the Thalamus.

PONS

Nucleus of Facial N. VII

Nucleus of Vagus Nerve X

MEDULLA

The 1ˢᵀ Sensory Neurones synapse with the 2ᴺᴰ Sensory Neurones in the Nuclei of Cranial Nerves V and VII in the Pons, and with X in the Medulla.

The cell bodies of the 1ˢᵀ Sensory Neurones lie in the Ganglia of Cranial Nerves V, VII and X

There is a wide link-up with lower centres in Spinal Cord.

PAIN and TEMPERATURE PATHWAYS from TRUNK and LIMBS

The nerve endings registering Pain and Warmth or Cold are linked by a chain of 3 Neurones with final Receiving Centre – the sensory area in the PARIETAL LOBES of the CEREBRAL CORTEX.

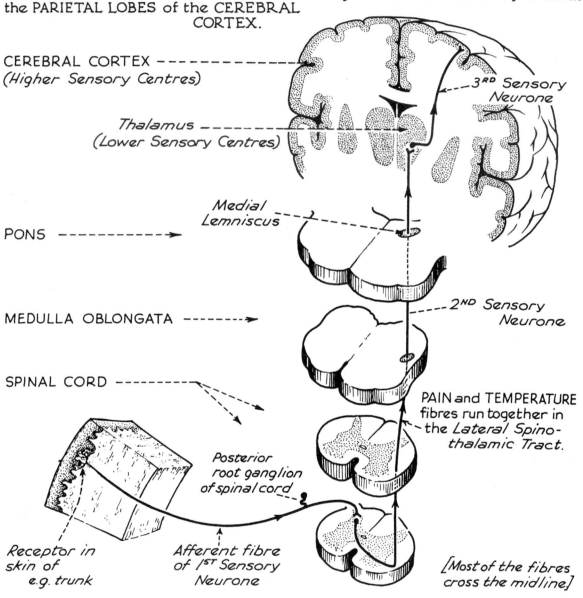

CEREBRAL CORTEX - - - - - - - - -
(Higher Sensory Centres)

3ᴿᴰ Sensory Neurone

Thalamus - - - - - - -
(Lower Sensory Centres)

Medial Lemniscus

PONS - - - - - - - - →

MEDULLA OBLONGATA - - - - - →

2ᴺᴰ Sensory Neurone

SPINAL CORD - - - - - →

PAIN and TEMPERATURE fibres run together in the *Lateral Spino-thalamic Tract.*

Posterior root ganglion of spinal cord

Receptor in skin of e.g. trunk

Afferent fibre of 1ˢᵀ Sensory Neurone

[Most of the fibres cross the midline]

TOUCH and PRESSURE PATHWAYS from TRUNK and LIMBS

Touch and Pressure endings are linked by a chain of 3 Neurones with the PARIETAL LOBES.

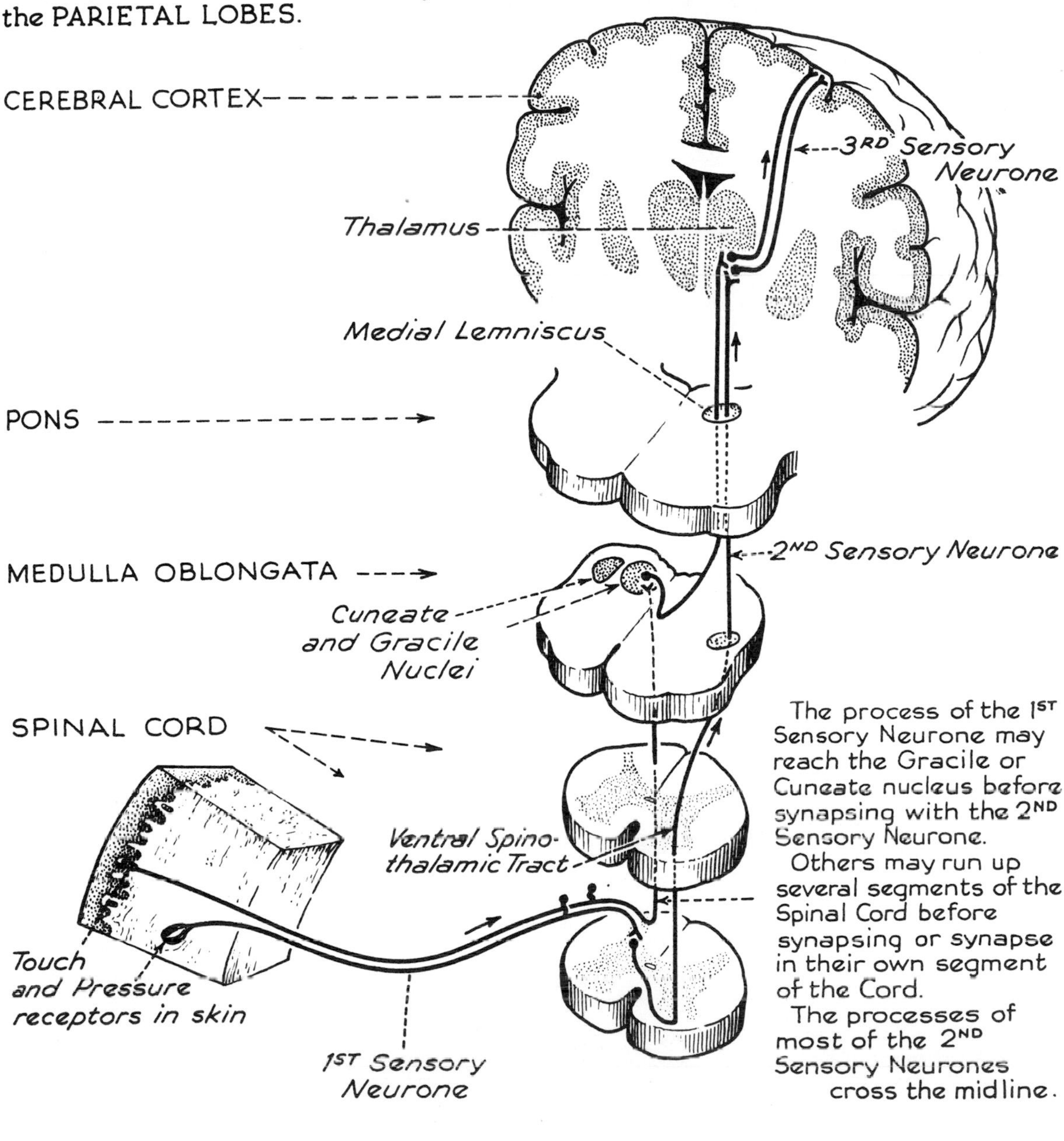

CEREBRAL CORTEX

---*3ᴿᴰ Sensory Neurone*

Thalamus

Medial Lemniscus

PONS

MEDULLA OBLONGATA

2ᴺᴰ Sensory Neurone

Cuneate and Gracile Nuclei

SPINAL CORD

The process of the 1ˢᵀ Sensory Neurone may reach the Gracile or Cuneate nucleus before synapsing with the 2ᴺᴰ Sensory Neurone.

Ventral Spino-thalamic Tract

Others may run up several segments of the Spinal Cord before synapsing or synapse in their own segment of the Cord.

The processes of most of the 2ᴺᴰ Sensory Neurones cross the midline.

Touch and Pressure receptors in skin

1ˢᵀ Sensory Neurone

249

SENSORY CORTEX

The 3ʳᴰ Sensory Neurones (conveying information from the <u>opposite side</u> of the body) synapse with cells in the POST-CENTRAL GYRUS of the PARIETAL LOBE of the CEREBRAL CORTEX. The exact points on this Gyrus at which impulses coming from the different regions of the skin surface terminate are indicated on this coronal view of the Gyrus

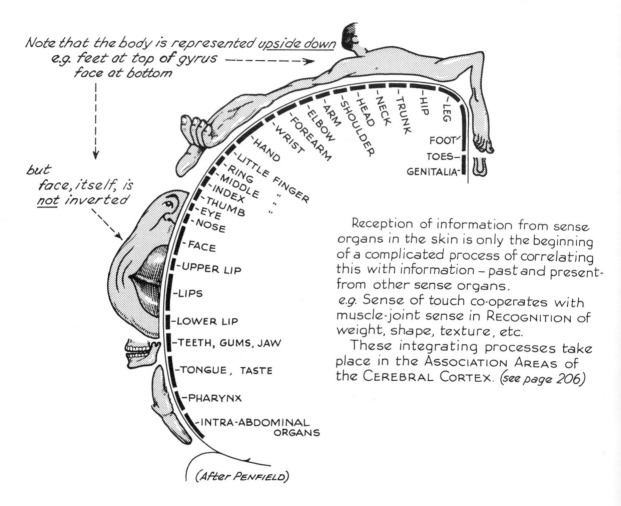

Note that the body is represented <u>upside down</u>
 e.g. feet at top of gyrus ------→
     face at bottom

but
face, itself, is
<u>not</u> inverted

LEG
HIP
TRUNK
NECK
HEAD
SHOULDER
ARM
ELBOW
FOREARM
WRIST
HAND
LITTLE FINGER
RING   "
MIDDLE "
INDEX   "
THUMB   "
EYE
NOSE
FOOT
TOES
GENITALIA

- FACE
- UPPER LIP
- LIPS
- LOWER LIP
- TEETH, GUMS, JAW
- TONGUE, TASTE
- PHARYNX
- INTRA-ABDOMINAL ORGANS

(After PENFIELD)

Reception of information from sense organs in the skin is only the beginning of a complicated process of correlating this with information – past and present – from other sense organs.
e.g. Sense of touch co-operates with muscle-joint sense in RECOGNITION of weight, shape, texture, etc.
These integrating processes take place in the ASSOCIATION AREAS of the CEREBRAL CORTEX. (see page 206)

Note the relatively large area devoted to FACE (especially lips) and to HAND (especially thumb and index finger) while trunk representation is very small.

MOTOR CORTEX

The MOTOR NERVE CELLS which send out impulses to initiate VOLUNTARY MOVEMENT of SKELETAL MUSCLES lie in the PRE-CENTRAL GYRUS of each FRONTAL LOBE in the CEREBRAL CORTEX.

One cerebral hemisphere controls the muscles on the opposite side of the body.

The exact point in the GYRUS where neurones controlling any one part of the body are situated is indicated in this coronal view of the Gyrus.

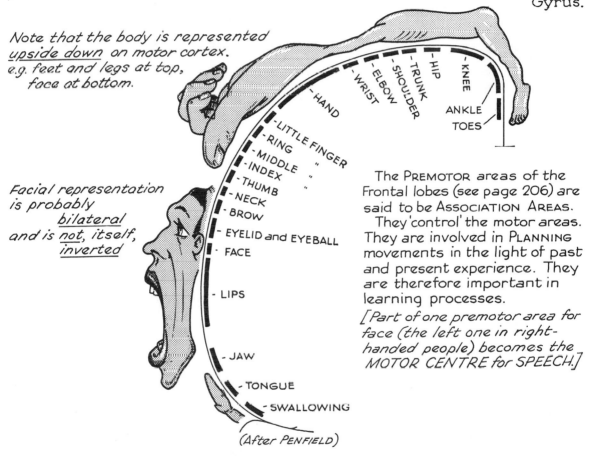

Note that the body is represented upside down on motor cortex. e.g. feet and legs at top, face at bottom.

Facial representation is probably bilateral and is not, itself, inverted

LITTLE FINGER
RING "
MIDDLE "
INDEX "
THUMB "
NECK
BROW
EYELID and EYEBALL
FACE
LIPS
JAW
TONGUE
SWALLOWING

HAND
WRIST
ELBOW
SHOULDER
TRUNK
HIP
KNEE
ANKLE
TOES

The PREMOTOR areas of the Frontal lobes (see page 206) are said to be ASSOCIATION AREAS.

They 'control' the motor areas. They are involved in PLANNING movements in the light of past and present experience. They are therefore important in learning processes.

[Part of one premotor area for face (the left one in right-handed people) becomes the MOTOR CENTRE for SPEECH.]

(After PENFIELD)

Note the large area of the Motor Cortex (and therefore the very large number of neurones) devoted to control of voluntary movements of the HANDS. This enables them to perform complicated movements and to acquire highly intricate skills: similarly with muscles involved in ARTICULATION.

MOTOR PATHWAYS to HEAD and NECK

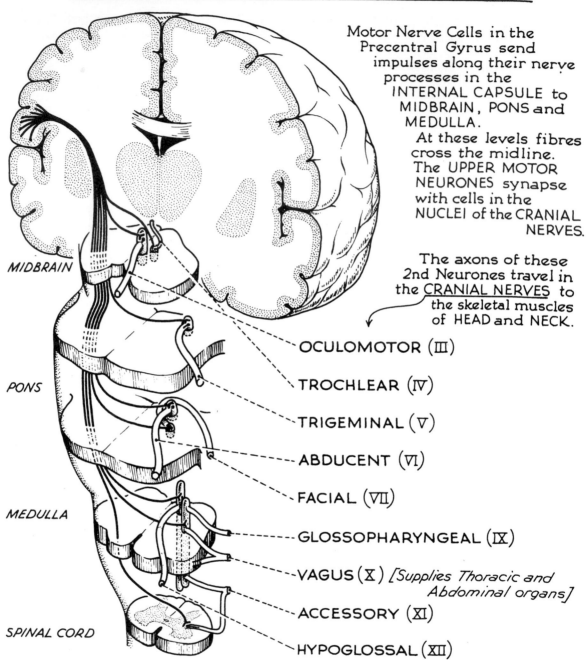

Motor Nerve Cells in the Precentral Gyrus send impulses along their nerve processes in the INTERNAL CAPSULE to MIDBRAIN, PONS and MEDULLA.

At these levels fibres cross the midline. The UPPER MOTOR NEURONES synapse with cells in the NUCLEI of the CRANIAL NERVES.

The axons of these 2nd Neurones travel in the CRANIAL NERVES to the skeletal muscles of HEAD and NECK.

MIDBRAIN

PONS

MEDULLA

SPINAL CORD

OCULOMOTOR (III)

TROCHLEAR (IV)

TRIGEMINAL (V)

ABDUCENT (VI)

FACIAL (VII)

GLOSSOPHARYNGEAL (IX)

VAGUS (X) [Supplies Thoracic and Abdominal organs]

ACCESSORY (XI)

HYPOGLOSSAL (XII)

MOTOR PATHWAYS to TRUNK and LIMBS

The CONTROLLING CENTRES in the MOTOR CORTEX are linked by 2 Neurones with the EFFECTOR ORGANS – the VOLUNTARY MUSCLES.

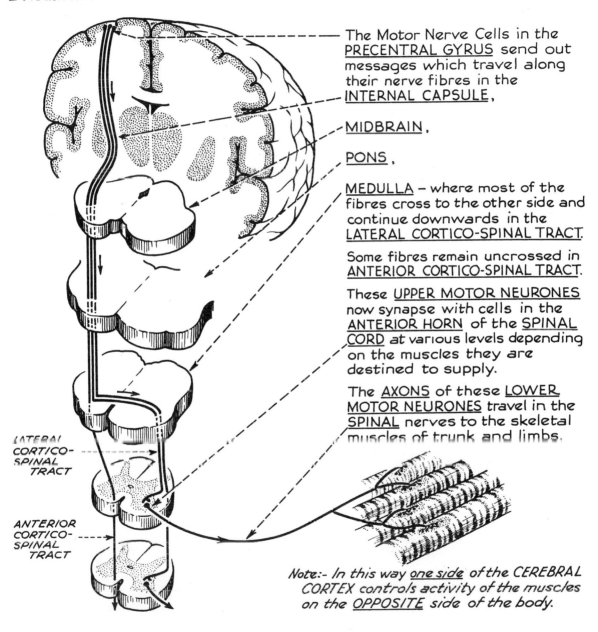

The Motor Nerve Cells in the PRECENTRAL GYRUS send out messages which travel along their nerve fibres in the INTERNAL CAPSULE,

MIDBRAIN,

PONS,

MEDULLA – where most of the fibres cross to the other side and continue downwards in the LATERAL CORTICO-SPINAL TRACT.

Some fibres remain uncrossed in ANTERIOR CORTICO-SPINAL TRACT.

These UPPER MOTOR NEURONES now synapse with cells in the ANTERIOR HORN of the SPINAL CORD at various levels depending on the muscles they are destined to supply.

The AXONS of these LOWER MOTOR NEURONES travel in the SPINAL nerves to the skeletal muscles of trunk and limbs.

LATERAL CORTICO-SPINAL TRACT

ANTERIOR CORTICO-SPINAL TRACT

Note:- In this way *one side* of the CEREBRAL CORTEX controls activity of the muscles on the *OPPOSITE* side of the body.

MOTOR UNIT

The AXON of the LOWER MOTOR NEURONE divides into many branches.
Each branch ends at the MOTOR END-PLATE of a single muscle fibre.

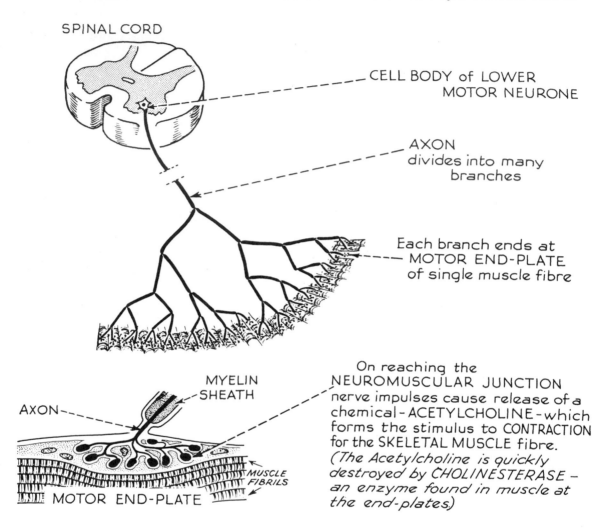

SPINAL CORD

CELL BODY of LOWER
MOTOR NEURONE

AXON
divides into many
branches

Each branch ends at
MOTOR END-PLATE
of single muscle fibre

MYELIN
SHEATH

AXON

On reaching the
NEUROMUSCULAR JUNCTION
nerve impulses cause release of a
chemical – ACETYLCHOLINE – which
forms the stimulus to CONTRACTION
for the SKELETAL MUSCLE fibre.
*(The Acetylcholine is quickly
destroyed by CHOLINESTERASE –
an enzyme found in muscle at
the end-plates.)*

MUSCLE
FIBRILS

MOTOR END-PLATE

*The motor nerve with the group of muscle fibres it supplies is
known as the MOTOR UNIT. (These muscle fibres may be widely
scattered within the anatomical muscle)*

FINAL COMMON PATHWAY

Each MOTOR NEURONE in the ANTERIOR HORNS of the SPINAL CORD serves as the PATHWAY for motor impulses initiated in —
<u>HIGHER MOTOR CENTRES in CEREBRUM</u> *(of opposite side)* and travelling in –

(1) *CORTICO-SPINAL TRACT*

The Motor Neurone also serves as the PATHWAY for co-ordinating corrective (restraining or facilitating) impulses discharged by—
<u>LOWER MOTOR CENTRES in EXTRAPYRAMIDAL SYSTEM</u> *(of same or opposite side)* and travelling in –

(2) *RUBRO-SPINAL TRACT*
from Red Nucleus *(opposite side)*

(3) *DORSAL VESTIBULO-SPINAL TRACT*
from Dorsal Vestibular Nucleus *(same side)*

(4) *OLIVO-SPINAL TRACT*
from Olivary Nucleus *(same side)*

(5) *RETICULO-SPINAL TRACT*
from Reticular Nuclei *(same side)*

(6) *VENTRAL VESTIBULO-SPINAL TRACT*
from Vestibular Nuclei *(opposite side)*

(7) *TECTO-SPINAL TRACT*
from Tectum *(opposite side)*

The Motor Neurone also receives relays of afferent impulses from other reflex centres in –
<u>SPINAL CORD</u>

(8) *For REFLEXES of <u>same</u> segment of cord and from <u>same</u> side of cord.*

(9) *For REFLEXES of same segment but <u>other</u> side of cord and body.*

(10) *For REFLEXES from <u>other</u> segments of cord but <u>same</u> side of cord.*

(11) *For REFLEXES from <u>other</u> segments and <u>other</u> side of cord.*

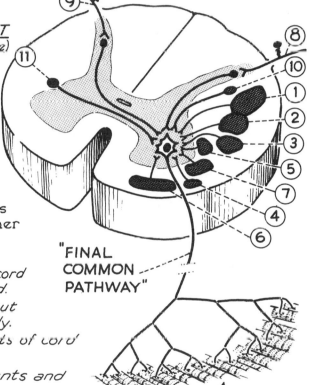

"FINAL COMMON PATHWAY"

Skeletal Muscle Fibres

If several sources compete for the "FINAL COMMON PATHWAY" at any one time, allied ones may reinforce each other: if incompatible, those which are initiated by PAINFUL stimuli (i.e. PROTECTIVE REFLEXES) take precedence.

EXTRAPYRAMIDAL SYSTEM

The actual performance of a VOLUNTARY MOVEMENT— initiated in the HIGHER MOTOR CENTRES in the CEREBRUM —involves co-operation by LOWER MOTOR CENTRES in the BRAIN STEM.

These Lower Centres are grouped together as the **EXTRAPYRAMIDAL SYSTEM**

CORPUS STRIATUM

CAUDATE NUCLEUS
LENTIFORM NUCLEUS:
    PUTAMEN,
    GLOBUS
    PALLIDUS

These centres have important 2-way connections with each other and with HIGHER and LOWER MOTOR and SENSORY Centres

CEREBELLUM
(for centres and connections see pp. 258-9)

MIDBRAIN and PONS
  RED NUCLEI
  SUBSTANTIA NIGRA
  RETICULAR NUCLEI
  (scattered groups of neurones)
  TECTUM
  VESTIBULAR NUCLEI

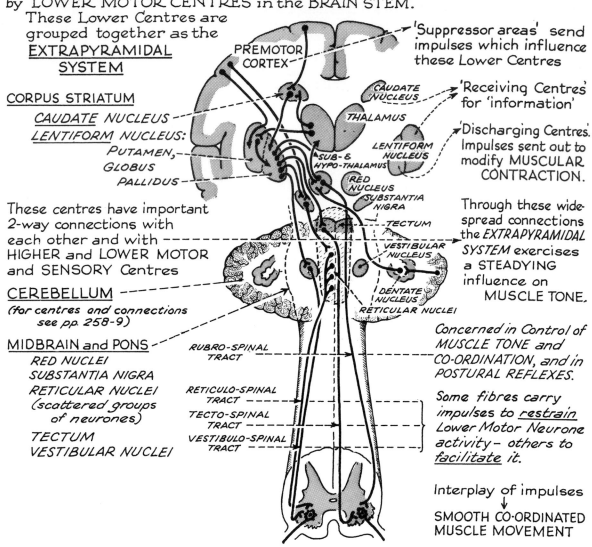

PREMOTOR CORTEX

CAUDATE NUCLEUS
THALAMUS
LENTIFORM NUCLEUS
SUB- & HYPO-THALAMUS
RED NUCLEUS
SUBSTANTIA NIGRA
TECTUM
VESTIBULAR NUCLEUS
DENTATE NUCLEUS
RETICULAR NUCLEI

RUBRO-SPINAL TRACT
RETICULO-SPINAL TRACT
TECTO-SPINAL TRACT
VESTIBULO-SPINAL TRACT

'Suppressor areas' send impulses which influence these Lower Centres

'Receiving Centres' for 'information'

'Discharging Centres.' Impulses sent out to modify MUSCULAR CONTRACTION.

Through these widespread connections the *EXTRAPYRAMIDAL SYSTEM* exercises a STEADYING influence on MUSCLE TONE.

Concerned in Control of MUSCLE TONE and CO-ORDINATION, and in POSTURAL REFLEXES.

Some fibres carry impulses to restrain Lower Motor Neurone activity — others to facilitate it.

Interplay of impulses
↓
SMOOTH CO-ORDINATED MUSCLE MOVEMENT

Little is known about the methods for regulating this interplay of impulses converging on the MOTOR UNIT during a voluntary action.

Where part of the Extrapyramidal System is destroyed by disease varying types of RIGIDITY, TREMOR and UNCO-ORDINATED MUSCLE MOVEMENT result.

CEREBELLUM

The CEREBELLUM has 2 HEMISPHERES.
Each Hemisphere has 3 LOBES. These differ in development and function.

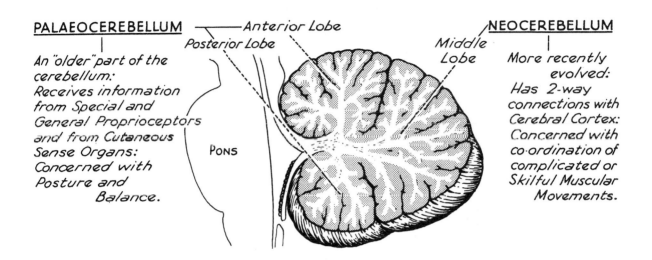

<u>PALAEOCEREBELLUM</u>

An "older" part of the cerebellum:
Receives information from Special and General Proprioceptors and from Cutaneous Sense Organs:
Concerned with Posture and Balance.

Anterior Lobe
Posterior Lobe

Pons

Middle Lobe

<u>NEOCEREBELLUM</u>

More recently evolved:
Has 2-way connections with Cerebral Cortex:
Concerned with co-ordination of complicated or Skilful Muscular Movements.

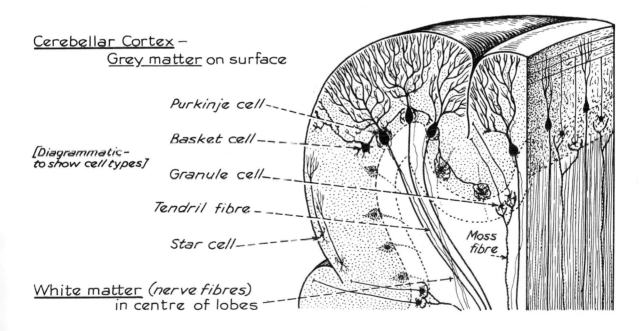

<u>Cerebellar Cortex</u> —
   <u>Grey matter</u> on surface

Purkinje cell

Basket cell

[Diagrammatic - to show cell types]

Granule cell

Tendril fibre

Star cell

Moss fibre

<u>White matter</u> *(nerve fibres)*
   in centre of lobes

CEREBELLUM

INGOING PATHWAYS

Each Cerebellar hemisphere is linked with the rest of the Nervous System through 3 bundles of nerve fibres — the Superior, Middle and Inferior Peduncles.

The Cerebellum receives information....

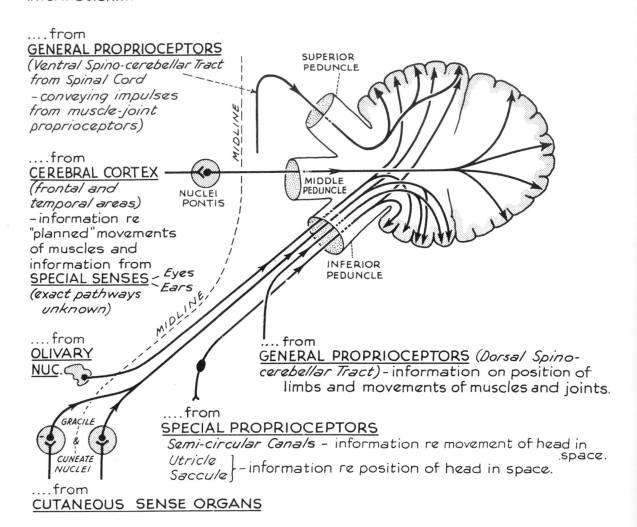

....from
GENERAL PROPRIOCEPTORS
(Ventral Spino-cerebellar Tract from Spinal Cord - conveying impulses from muscle-joint proprioceptors)

....from
CEREBRAL CORTEX
(frontal and temporal areas) - information re "planned" movements of muscles and information from
SPECIAL SENSES - *Eyes* *Ears*
(exact pathways unknown)

....from
OLIVARY NUC.

SUPERIOR PEDUNCLE

MIDLINE

NUCLEI PONTIS

MIDDLE PEDUNCLE

INFERIOR PEDUNCLE

MIDLINE

GRACILE
&
CUNEATE NUCLEI

....from
GENERAL PROPRIOCEPTORS *(Dorsal Spino-cerebellar Tract)* - information on position of limbs and movements of muscles and joints.

....from
SPECIAL PROPRIOCEPTORS
Semi-circular Canals - information re movement of head in space.
Utricle *Saccule* } - information re position of head in space.

....from
CUTANEOUS SENSE ORGANS

The activities of the Cerebellum are carried out beneath the level of consciousness. The information reaching it gives no sensation.

CEREBELLUM

OUTGOING PATHWAYS

In addition to the grey matter on the surface of the cerebellar cortex, there are masses of grey matter — called CEREBELLAR NUCLEI — deep within the cerebellum.

The **PALAEOCEREBELLUM**
sends out messages to...
..... RED NUCLEUS ——
of opposite side to influence
the activity of the primitive
Rubro-spinal motor pathway.

..... CRANIAL NERVE NUCLEI
III, IV, VI and XI to co-ordinate
movements and position of
head with movements of
eyes and trunk.

Vestibular Nucleus

Vestibulo-spinal tract
to co-ordinate movements
and position of head with
Postural Tone of limb muscles.

The **NEOCEREBELLUM**
sends out messages to.....
....PRE-MOTOR region of
CEREBRAL CORTEX (via
THALAMUS of opposite
side) to influence
"Voluntary" actions.

.....RED NUCLEUS of
opposite side to influence
activity of the Rubro-
spinal motor pathway

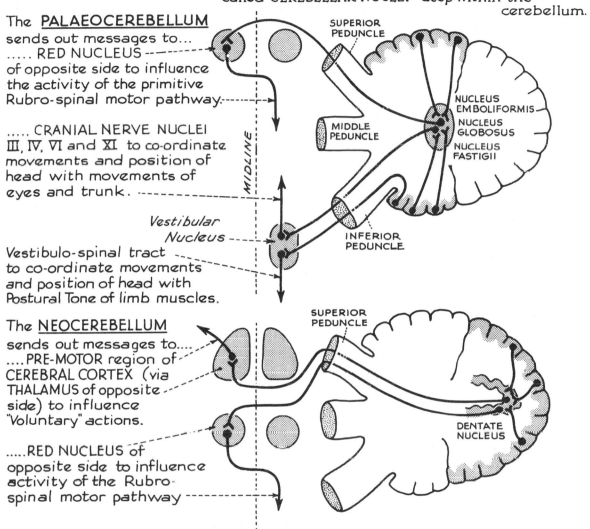

SUPERIOR PEDUNCLE

MIDDLE PEDUNCLE

MIDLINE

NUCLEUS EMBOLIFORMIS
NUCLEUS GLOBOSUS
NUCLEUS FASTIGII

INFERIOR PEDUNCLE

SUPERIOR PEDUNCLE

DENTATE NUCLEUS

The cerebral cortex initiates purposeful movements. During such movements proprioceptors are continually supplying information to the Cerebellum about the changing positions of muscles and joints. The Cerebellum is then responsible for collating this information and sending out impulses to bring about the smooth co-ordinated movements of different groups of muscles.

259

CONTROL of MUSCLE MOVEMENT

The muscle spindle is the key structure in the complex self-regulating mechanism for the control of movement of skeletal muscle.

In VOLUNTARY MOVEMENT a muscle can be made to contract by impulses reaching it by one of 2 routes:-

DIRECT PATHWAY

Impulses from HIGHER CENTRES in CEREBRAL CORTEX pass down large motor neurones of the Pyramidal tract — the alpha (α) route — to excite the LOWER MOTOR NEURONES → MOTOR UNITS — and lead to CONTRACTION of MUSCLE.

– This is the pathway involved in *voluntary movements*

INDIRECT PATHWAY

Impulses from HIGHER CENTRES especially the CEREBELLUM pass in small motor neurones to intrafusal muscle fibres in the MUSCLE SPINDLE — the gamma (γ) route — These contract and cause stretching of spiral sensory endings which then discharge an increased number of impulses along their afferent fibres into the Spinal Cord.
This reflexly leads to the discharge of impulses in the LOWER MOTOR NEURONE and CONTRACTION of MUSCLE.

– This pathway (in conjunction with the direct route) is necessary for *accurately controlled movements* and also for movements involved in *maintenance of posture.*

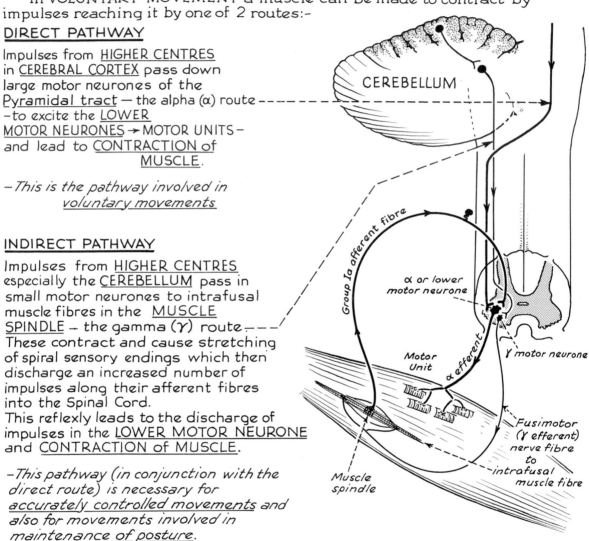

It is probable that the CEREBRAL CORTEX decides *WHAT* is to be done and the CEREBELLUM 'decides' *HOW* it is to be done.

LOCOMOTOR SYSTEM

BONES and MUSCLES ———— concerned with MOVEMENT of the body.

SKELETON——RIGID FRAMEWORK gives SHAPE and SUPPORT to body.

is JOINTED to permit MOVEMENT

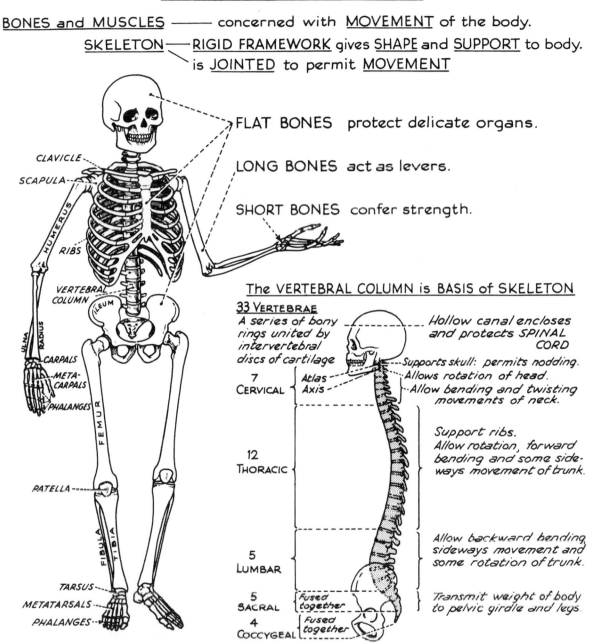

FLAT BONES protect delicate organs.

LONG BONES act as levers.

SHORT BONES confer strength.

CLAVICLE

SCAPULA

HUMERUS

RIBS

VERTEBRAL COLUMN

ILEUM

ULNA

RADIUS

CARPALS

META-CARPALS

PHALANGES

FEMUR

PATELLA

FIBULA

TIBIA

TARSUS

METATARSALS

PHALANGES

The VERTEBRAL COLUMN is BASIS of SKELETON

33 VERTEBRAE

A series of bony rings united by intervertebral discs of cartilage

Hollow canal encloses and protects SPINAL CORD

7 CERVICAL	Atlas	Supports skull: permits nodding.
	Axis	Allows rotation of head.
		Allow bending and twisting movements of neck.

Support ribs.
Allow rotation, forward bending and some sideways movement of trunk.

12 THORACIC

Allow backward bending, sideways movement and some rotation of trunk.

5 LUMBAR

5 SACRAL — Fused together

Transmit weight of body to pelvic girdle and legs.

4 COCCYGEAL — Fused together

All bones give attachment to muscles.

SKELETAL MUSCLES

Bones are moved at joints by the CONTRACTION and RELAXATION of MUSCLES attached to them.

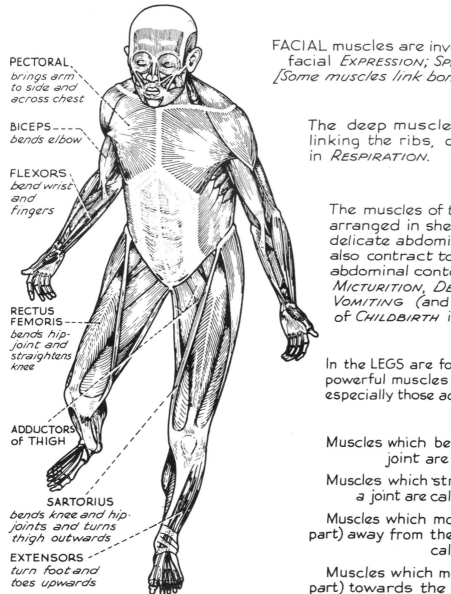

PECTORAL,
brings arm
to side and
across chest

BICEPS
bends elbow

FLEXORS
bend wrist
and
fingers

RECTUS
FEMORIS
bends hip-
joint and
straightens
knee

ADDUCTORS
of THIGH

SARTORIUS
bends knee and hip-
joints and turns
thigh outwards

EXTENSORS
turn foot and
toes upwards

FACIAL muscles are involved in varying facial *EXPRESSION*; *SPEECH*; *MASTICATION* [*Some muscles link bone to skin.*]

The deep muscles of the THORAX, linking the ribs, contract and relax in *RESPIRATION*.

The muscles of the ABDOMEN are arranged in sheets and *PROTECT* delicate abdominal organs. They also contract to compress abdominal contents and aid in *MICTURITION*, *DEFAECATION*, *VOMITING* (and in the processes of *CHILDBIRTH* in the female).

In the LEGS are found the most powerful muscles of the body – especially those acting on the hip-joint.

Muscles which bend a limb at a joint are called **FLEXORS**.

Muscles which straighten a limb at a joint are called **EXTENSORS**.

Muscles which move a limb (or other part) away from the midline are called **ABDUCTORS**.

Muscles which move a limb (or other part) towards the midline are called **ADDUCTORS**.

262

SKELETAL MUSCLES

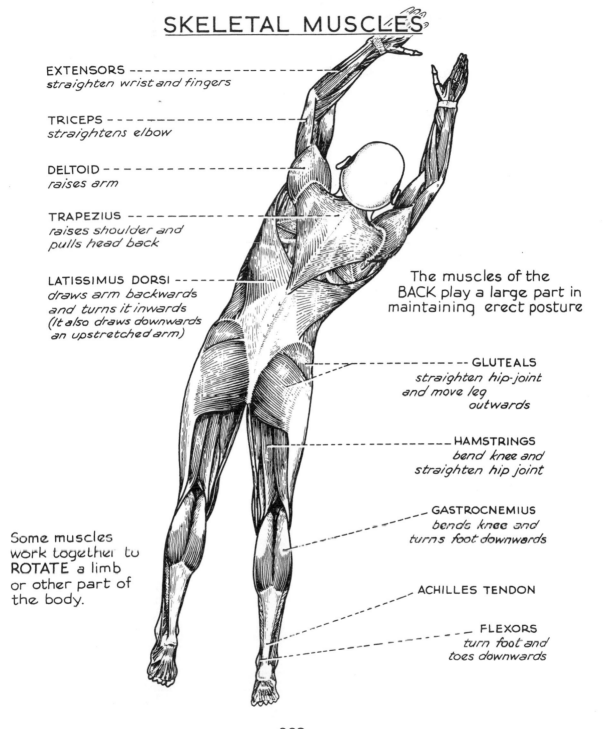

EXTENSORS
straighten wrist and fingers

TRICEPS
straightens elbow

DELTOID
raises arm

TRAPEZIUS
*raises shoulder and
pulls head back*

LATISSIMUS DORSI
*draws arm backwards
and turns it inwards
(It also draws downwards
an upstretched arm)*

The muscles of the
BACK play a large part in
maintaining erect posture

GLUTEALS
*straighten hip-joint
and move leg
outwards*

HAMSTRINGS
*bend knee and
straighten hip joint*

GASTROCNEMIUS
*bends knee and
turns foot downwards*

ACHILLES TENDON

FLEXORS
*turn foot and
toes downwards*

Some muscles
work together to
ROTATE a limb
or other part of
the body.

263

MUSCULAR MOVEMENTS

The long bones particularly form a light framework of LEVERS.
The skeletal muscles attached to them contract to operate these levers.

When a muscle contracts it shortens.

This brings its two ends closer together.

Since the two ends are attached to different bones by TENDONS one or other of the bones must move.

Two bones meet or articulate at a JOINT.

Joint surfaces are covered with a layer of smooth CARTILAGE.

To avoid friction when the two surfaces move on one another a SYNOVIAL MEMBRANE secretes a LUBRICATING FLUID.

BICEPS

TRICEPS

FLEXION

EXTENSION

The muscle which contracts to move the joint is called the PRIME MOVER or AGONIST *(the Biceps in flexion of elbow)*

To allow the movement to take place, however, other muscles near the joint must co-operate :-
The oppositely acting muscles gradually relax — these are called the ANTAGONISTS and exercise a "braking" control on the movement. *(e.g. the EXTENSORS — chiefly Triceps — in flexion of the elbow.)*
Other muscles steady the bone giving "origin" to the Prime Mover so that only the "insertion" will move — these muscles are called FIXATORS.
Still other muscles help to steady, for most efficient movement, the joint being moved — called SYNERGISTS.

When the elbow is straightened the reverse occurs :-
TRICEPS, the Prime Mover, CONTRACTS; BICEPS, the Antagonist, RELAXES.

RECIPROCAL INNERVATION

The co-ordinated group action of muscles is made possible by the many synaptic connections between CONNECTOR NEURONES of the INGOING or PROPRIOCEPTIVE NEURONES of one muscle group and the OUTGOING or MOTOR NEURONES of the functionally opposite group of muscles.

This is shown diagrammatically for the alternating contractions and relaxations of the Flexors and Extensors of one knee in Walking:-

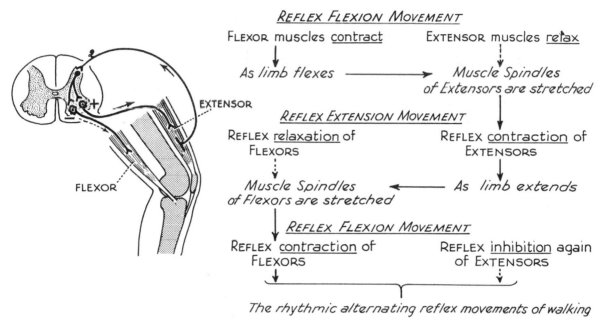

REFLEX FLEXION MOVEMENT

FLEXOR muscles contract   EXTENSOR muscles relax

As limb flexes ⟶ Muscle Spindles of Extensors are stretched

REFLEX EXTENSION MOVEMENT

REFLEX relaxation of FLEXORS   REFLEX contraction of EXTENSORS

Muscle Spindles of Flexors are stretched ⟵ As limb extends

REFLEX FLEXION MOVEMENT

REFLEX contraction of FLEXORS   REFLEX inhibition again of EXTENSORS

The rhythmic alternating reflex movements of walking

i.e. The FINAL COMMON PATHWAY *is commandeered alternately by "exciting" or "inhibiting" messages.*

RECIPROCAL INNERVATION is probably due to division (within the SPINAL CORD) of the CONNECTOR NEURONES for each AFFERENT NERVE – e.g. one branch excites (+) extensor motor neurone; the other branch inhibits (–) the flexor motor neurone.

Thus:–

When limb is flexed–

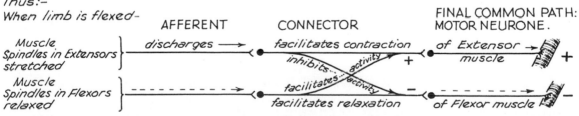

	AFFERENT	CONNECTOR	FINAL COMMON PATH: MOTOR NEURONE.
Muscle Spindles in Extensors stretched	discharges ⟶	facilitates contraction / inhibits...activity +	of Extensor muscle ⟶ +
Muscle Spindles in Flexors relaxed	- - - - - ⟶	facilitates activity / facilitates relaxation –	of Flexor muscle –

RECIPROCAL INNERVATION helps make possible self-regulation of the rhythmical movements.

SKELETAL MUSCLE
and the MECHANISM of CONTRACTION

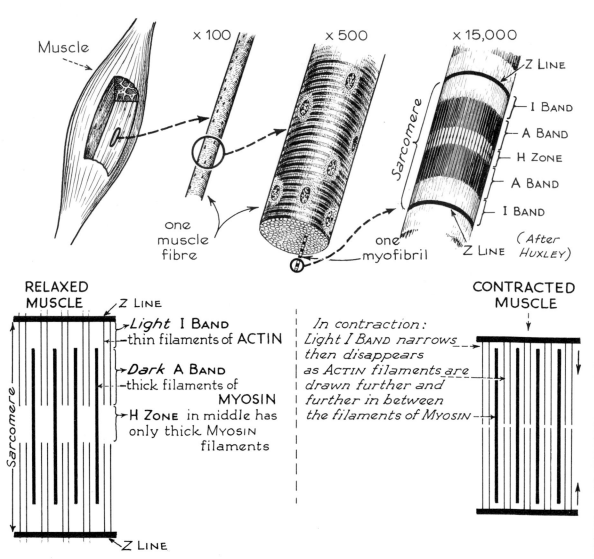

Muscle

× 100

× 500

× 15,000

Z LINE
I BAND
A BAND
H ZONE
A BAND
I BAND

Sarcomere

one muscle fibre

one myofibril

Z LINE

(After HUXLEY)

RELAXED MUSCLE

Z LINE

Light I BAND
—thin filaments of ACTIN

Dark A BAND
—thick filaments of MYOSIN

H ZONE in middle has only thick MYOSIN filaments

Sarcomere

Z LINE

In contraction:
Light I BAND narrows then disappears as ACTIN filaments are drawn further and further in between the filaments of MYOSIN

CONTRACTED MUSCLE

Energy for contraction is derived from glucose and fat in the mitochondrion (*page 6*). The energy is transported from the mitochondrion to the contractile filaments in ATP. ATP splits readily into ADP and phosphate, releasing its trapped energy where needed (*page 42*).

266

AUTONOMIC NERVOUS SYSTEM

and

CHEMICAL TRANSMISSION at NERVE ENDINGS

AUTONOMIC NERVOUS SYSTEM

The Autonomic Nervous System is concerned with maintaining a <u>Stable Internal Environment</u>. It governs functions which are normally carried out below the level of consciousness.

It has 2 separate parts – <u>PARASYMPATHETIC</u> and <u>SYMPATHETIC</u>. Both have their Highest Centres in the Brain.

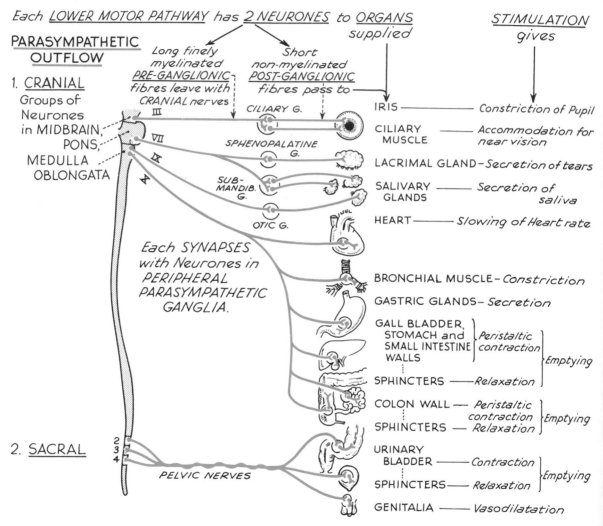

Each <u>LOWER MOTOR PATHWAY</u> has <u>2 NEURONES</u> to <u>ORGANS</u> supplied

<u>STIMULATION</u> *gives*

PARASYMPATHETIC OUTFLOW

Long finely myelinated <u>PRE-GANGLIONIC</u> *fibres leave with CRANIAL nerves*

Short non-myelinated <u>POST-GANGLIONIC</u> *fibres pass to*

1. <u>CRANIAL</u>
Groups of Neurones in MIDBRAIN, PONS, MEDULLA OBLONGATA

III — CILIARY G.
VII — SPHENOPALATINE G.
IX — SUB-MANDIB. G.
X — OTIC G.

Each SYNAPSES with Neurones in PERIPHERAL PARASYMPATHETIC GANGLIA.

IRIS ———————— *Constriction of Pupil*

CILIARY MUSCLE ——— *Accommodation for near vision*

LACRIMAL GLAND – *Secretion of tears*

SALIVARY GLANDS ——— *Secretion of saliva*

HEART ———————— *Slowing of Heart rate*

BRONCHIAL MUSCLE – *Constriction*

GASTRIC GLANDS – *Secretion*

GALL BLADDER, STOMACH and SMALL INTESTINE WALLS } *Peristaltic contraction*

SPHINCTERS ——— *Relaxation*

} *Emptying*

COLON WALL —— *Peristaltic contraction*

SPHINCTERS —— *Relaxation*

} *Emptying*

2. <u>SACRAL</u>
2
3
4
PELVIC NERVES

URINARY BLADDER ——— *Contraction*

SPHINCTERS ——— *Relaxation*

} *Emptying*

GENITALIA ——— *Vasodilatation*

General stimulation of the Parasympathetic promotes vegetative functions of the body. Stimulation of the Parasympathetic and inhibition of Sympathetic have the same overall effects.

AUTONOMIC NERVOUS SYSTEM

The Parasympathetic and Sympathetic systems normally act in balanced reciprocal fashion.   The activity of an organ at any one time is the result of the two opposing influences.

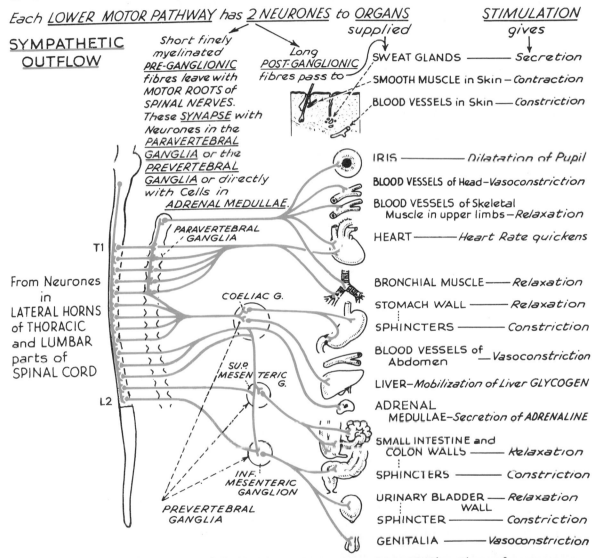

Each _LOWER MOTOR PATHWAY_ has _2 NEURONES_ to ORGANS supplied   _STIMULATION_ gives

SYMPATHETIC OUTFLOW

Short finely myelinated _PRE-GANGLIONIC_ fibres leave with MOTOR ROOTS of SPINAL NERVES. These _SYNAPSE_ with Neurones in the _PARAVERTEBRAL GANGLIA_ or the _PREVERTEBRAL GANGLIA_ or directly with Cells in _ADRENAL MEDULLAE._

Long _POST-GANGLIONIC_ fibres pass to

SWEAT GLANDS ———— Secretion
SMOOTH MUSCLE in Skin — Contraction
BLOOD VESSELS in Skin —— Constriction

IRIS ———— Dilatation of Pupil
BLOOD VESSELS of Head – Vasoconstriction
BLOOD VESSELS of Skeletal Muscle in upper limbs – Relaxation
HEART ———— Heart Rate quickens

BRONCHIAL MUSCLE ——— Relaxation
STOMACH WALL ——— Relaxation
SPHINCTERS ———— Constriction
BLOOD VESSELS of Abdomen —— Vasoconstriction
LIVER – Mobilization of Liver GLYCOGEN
ADRENAL MEDULLAE – Secretion of ADRENALINE
SMALL INTESTINE and COLON WALLS ——— Relaxation
SPHINCTERS ———— Constriction
URINARY BLADDER — Relaxation WALL
SPHINCTER ———— Constriction
GENITALIA ———— Vasoconstriction

PARAVERTEBRAL GANGLIA

T1

From Neurones in LATERAL HORNS of THORACIC and LUMBAR parts of SPINAL CORD

COELIAC G.

SUP. MESENTERIC G.

L2

INF. MESENTERIC GANGLION

PREVERTEBRAL GANGLIA

General stimulation of Sympathetic results in mobilization of resources to prepare body to meet emergencies.   Stimulation of the Sympathetic and inhibition of the Parasympathetic have the same overall effects.

AUTONOMIC REFLEX

AUTONOMIC CENTRES in the BRAIN and SPINAL CORD receive SENSORY INFLOWS from the VISCERA. (Less is known about their exact pathways than about MOTOR OUTFLOWS.) Some of the SENSORY NEURONES convey information about events in the viscera to HIGHER AUTONOMIC CENTRES which send impulses to modify the activity of ⎯⎯⎯⎯⎯⎯⎯⎯

Both VISCERAL and SOMATIC AFFERENTS serve as AFFERENT PATHWAYS for AUTONOMIC REFLEXES by means of which much of the nervous regulation of vegetative functions is carried out below the level of consciousness.

e.g. The Simplest Autonomic Reflex arc:-

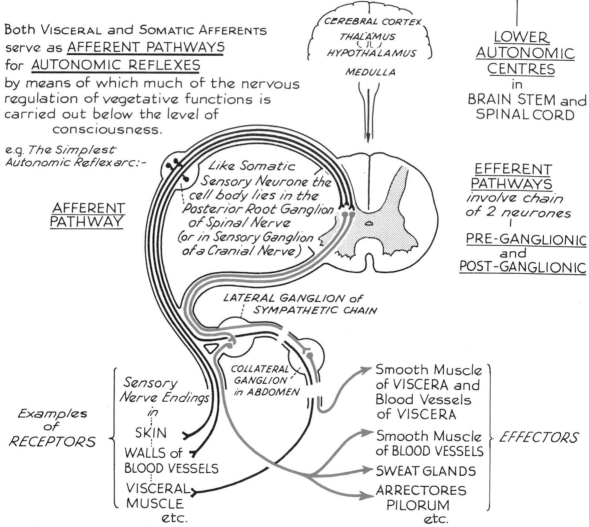

CEREBRAL CORTEX
THALAMUS
HYPOTHALAMUS
MEDULLA

LOWER AUTONOMIC CENTRES in BRAIN STEM and SPINAL CORD

AFFERENT PATHWAY

Like Somatic Sensory Neurone the cell body lies in the Posterior Root Ganglion of Spinal Nerve (or in Sensory Ganglion of a Cranial Nerve)

EFFERENT PATHWAYS involve chain of 2 neurones

PRE-GANGLIONIC and POST-GANGLIONIC

LATERAL GANGLION of SYMPATHETIC CHAIN

COLLATERAL GANGLION in ABDOMEN

Examples of RECEPTORS

Sensory Nerve Endings in SKIN WALLS of BLOOD VESSELS VISCERAL MUSCLE etc.

Smooth Muscle of VISCERA and Blood Vessels of VISCERA

Smooth Muscle of BLOOD VESSELS

SWEAT GLANDS

ARRECTORES PILORUM etc.

EFFECTORS

The Autonomic Reflex Arc differs from the Somatic Reflex Arc mainly in that it has 2 Efferent neurones. Transmission of impulse from Afferent to Efferent probably involves one or more connecting neurones.

CHEMICAL TRANSMISSION at NERVE ENDINGS

When an impulse passes along a nerve, sodium enters the fibre and potassium leaves it. When this process reaches the synapse between neurone and neurone or between neurone and effector organ a chemical substance is liberated. This bridges the gap and forms the stimulus to alter the membrane potential and initiate in the next neurone (or in the effector) the wave of depolarization and altered membrane permeability with ion exchange which becomes the transmitted nerve impulse.

Some nerves when stimulated liberate an acetylcholine-like substance. These are called CHOLINERGIC. Others liberate nor-adrenaline. These are called ADRENERGIC.

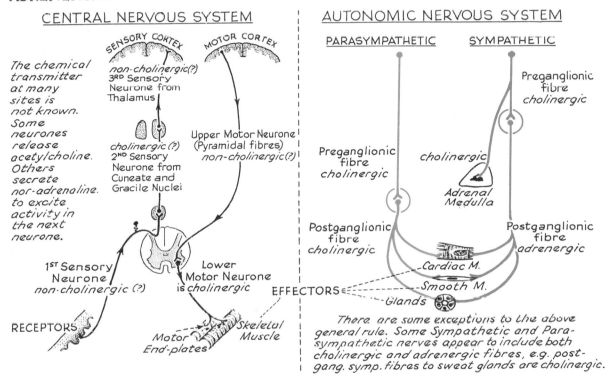

CENTRAL NERVOUS SYSTEM

SENSORY CORTEX   MOTOR CORTEX

The chemical transmitter at many sites is not known. Some neurones release acetylcholine. Others secrete nor-adrenaline. to excite activity in the next neurone.

non-cholinergic(?)
3RD Sensory Neurone from Thalamus

cholinergic (?)
2ND Sensory Neurone from Cuneate and Gracile Nuclei

Upper Motor Neurone (Pyramidal fibres)
non-cholinergic(?)

1ST Sensory Neurone
non-cholinergic (?)

Lower Motor Neurone is cholinergic

RECEPTORS

Motor End-plates   Skeletal Muscle

AUTONOMIC NERVOUS SYSTEM

PARASYMPATHETIC   SYMPATHETIC

Preganglionic fibre cholinergic

Preganglionic fibre cholinergic

cholinergic

Adrenal Medulla

Postganglionic fibre cholinergic

Postganglionic fibre adrenergic

Cardiac M.
Smooth M.
Glands

EFFECTORS

There are some exceptions to the above general rule. Some Sympathetic and Parasympathetic nerves appear to include both cholinergic and adrenergic fibres, e.g. postgang. symp. fibres to sweat glands are cholinergic.

There are thought to be other chemical substances secreted by neurones which may act as Synaptic Transmitters to regulate certain aspects of cerebral activity.
Many neurones within the C.N.S. release inhibitory, _not_ excitatory, transmitters. One such substance found in the cerebral cortex is GABA (gamma aminobutyric acid). Another is the amino acid, glycine.
Inhibitory neurones block unimportant signals and thus permit onward transmission of selected information e.g. to centres which otherwise would be subjected to continuous bombardment by impulses from receptors all over the body.

INDEX

273

INDEX